AF530252

Principles of Crop Improvement
Objective Analysis

The Author

Dr. Phundan Singh, hails from a reputed agricultural family of Western Uttar Pradesh [Village-Sakauti, Tehsil- Mawana, District- Meerut]. He graduated in Agriculture from Agra University, Agra and obtained his Master and Doctorate degrees from Kanpur University, Kanpur. He has throughout a brilliant academic record. His area of specialization is Plant Breeding and Genetics. Dr. P. Singh has over 40 year experience in the field of Plant Breeding and Genetics. He has served the Central Institute for Cotton Research, Nagpur for about 32 years in various capacities such as Scientist, Senior Scientist, Principal Scientist, Head of Division and Director. He has 300 publications to his credit. He has participated and presented 80 research papers in various National and International Seminars, Symposia and Conferences. He has authored more than 80 books related to Plant Breeding and Genetics. He has also authored 18 technical bulletins on various aspects of cotton and contributed 14 chapters in various books. Dr P. Singh is a member of several Scientific Societies, referee of different journals, Expert member of Selection Committees and M. Sc. and Ph. D. Examiner in different Agricultural Universities. The Institute was recognized as the referral Laboratory for detection of Bt. gene and received the best Annual Report Award in 2003 when he was the Director. He was conferred with Lifetime achievement Award in 2018 by Cotton Research and Development Association, Hisar. He has visited several countries such as Russia, Belarus, Ukraine, Crimea, Uzbekistan, Canada, USA, England, Austria and Germany.

Principles of Crop Improvement
Objective Analysis

[As per 5th Deans' Committee of ICAR]

— Author—

PHUNDAN SINGH

Former Director
ICAR-Central Institute for Cotton Research,
Nagpur – 440 010
Maharashtra, India

2025

Astral Cracker

A Division of

Astral International Pvt. Ltd.

New Delhi – 110 002

© 2025 AUTHOR
ISBN: 978-93-5919-822-4

Publisher's Note:

Every possible effort has been made to ensure that the information contained in this book is accurate at the time of going to press, and the publisher and author cannot accept responsibility for any errors or omissions, however caused. No responsibility for loss or damage occasioned to any person acting, or refraining from action, as a result of the material in this publication can be accepted by the editor, the publisher or the author. The Publisher is not associated with any product or vendor mentioned in the book. The contents of this work are intended to further general scientific research, understanding and discussion only. Readers should consult with a specialist where appropriate.

Every effort has been made to trace the owners of copyright material used in this book, if any. The author and the publisher will be grateful for any omission brought to their notice for acknowledgment in the future editions of the book.

All Rights reserved under International Copyright Conventions. No part of this publication may be reproduced, stored in a retrieval system, or transmitted in any form or by any means, electronic, mechanical, photocopying, recording or otherwise without the prior written consent of the publisher and the copyright owner.

Published by : **Astral Cracker**
A Division of
Astral International Pvt. Ltd.
– ISO 9001:2015 Certified Company –
4736/23, Ansari Road, Darya Ganj,
New Delhi-110 002
Ph. 011-43549197, 23278134
E-mail: info@astralint.com
Website: www.astralint.com

— Dedicated to —

Dr. Saksham Singh [Daughter-in-law]
Ms. Subhika Singh [Granddaughter]
Mr. Rishabh Singh [Grandson]

Preface

Plant breeding plays an important role in genetic improvement of crop plants in relation to their economic use for human being. As per 5th Dean Committee Report of ICAR, a new paper entitled "Crop Improvement" has been introduced. There are several books of Plant Breeding. However, there is hardly any book which covers the syllabus of under graduate course related to Crop Improvement in objective manner. The objective of writing the present book is to fulfill this long felt need of Indian students.

This book has been designed to covers the undergraduate syllabus of Crop Improvement as per ICAR 5th Dean committee report, 2016. It covers the entire syllabus in one compact volume of 20 chapters. Glossary of technical terms is presented at the end. Some useful information has been appended for general knowledge of students.

The book has been divided into four sections. Section I deals with Introductory Topics, section II with Breeding Methods, III with Seed Production and Crop Breeding and Section IV with Miscellaneous Topics. The book has been written in a simple language for easy grasping of undergraduate students.

The information contained in this book has been gathered from various published sources and Internet websites. Attempts have been made to provide latest information even then some valuable information might have been missed.

The cooperation extended by my wife Mrs. Jaswanti Singh during preparation of this book is praiseworthy. Hope, this Volume would be useful to the students, researchers, and teachers engaged in the field of Crop Improvement. Constructive suggestions of students and teachers are invited for further improvement of this book.

Phundan Singh

Contents

SECTION III: SEED PRODUCTION AND CROP BREEDING

SECTION IV: MISCELLANEOUS TOPICS

Syllabus of Crop Improvement I and Crop Improvement II

[As per 5th Deans' Committee of ICAR]

1. Centers of origin, distribution of species, wild relatives in different cereals; pulses; oilseeds; fibres; fodders and cash crops; vegetable and horticultural crops;
2. Plant genetic resources, its utilization and conservation,
3. Study of genetics of qualitative and quantitative characters;
4. Important concepts of breeding self-pollinated, cross-pollinated and vegetatively propagated crops;
5. Major breeding objectives and procedures including conventional and modern innovative approaches for development of hybrids and varieties for yield, adaptability, stability, abiotic and biotic stress tolerance and quality (physical, chemical, nutritional);
6. Hybrid seed production technology of Kharif Crops [Maize, Rice, Sorghum, Pearl millet and Pigeon pea, *etc.*],
7. Hybrid seed production of rabi crops
8. Ideotype concept and
9. Climate resilient crop varieties for future.

Section I

Introductory Topics

Chapter 1

Introduction and Objectives

A. DEFINITIONS

Abiotic Stress: Adverse conditions imposed for growth and production of crop plants by environmental factors such as deficiency or excess of nutrition, moisture, temperature and light; presence of harmful gases or toxicants; and abnormal soil conditions, such as salinity, alkalinity and acidity.

Adaptability: Stable performance of a variety over a wide range of environmental conditions.

Biotic Stress: Adverse conditions imposed for growth and production of crop plants by biotic factors such as insects, diseases and parasitic weeds.

Breeding techniques: Various breeding procedures which are used for genetic improvement of crop plants in relation to their economic use.

Easy Care Property: Cotton fabrics that are easily washed and require little ironing or pressing.

Genetic Resistance: Ability of some genotypes to give higher yield of good quality than other varieties at the same initial level of disease or insect infestation under similar environmental conditions.

Germplasm: The whole library of alleles in a crop species or sum total of genes in a species.

Harvest Index: The ratio of economic yield to biological yield. Biological yield is the total dry matter production per plant.

Plant Breeding: A science, an art and a technology which deals with genetic improvement of crop plants in relation to their economic use for mankind, also called as crop improvement.

Quality: Suitability or fitness of an economic plant product in relation to its end use.

Seed Production Technology: A branch of plant breeding which deals with principles and methods of improved seed production.

B. BACKGROUND INFORMATION

1. Plant breeding is an applied branch of plant genetics which has wide practical application in crop improvement.
2. In 1979, N. W. Simmonds has defined plant breeding as the current phase of crop evolution.
3. The history of plant breeding can be broadly divided in to two parts, *viz.* pre-Mendelian era and post Mendelian era.
4. Pre-Mendelian era started from the beginning of 18th century.
5. As early as in 1717, Thomas Fairchild developed first interspecific hybrid between sweet William and carnation species of Dianthus (Dianthus barbatus x D. Caryophyllus).
6. Later on in 1800, Thomas Andrew Knight First used artificial hybridization in fruit crops.
7. In 1840, Jonn Le Couteur and Partyic Shireff developed the concept of progeny test and individual plant selection in cereals.
8. In plant breeding, there are two types of objectives, *viz.*, general objectives and crop specific objectives. Major objectives of plant breeding are improvement in yield, quality, resistance to biotic and abiotic stresses, earliness, toxin free products, *etc.*
9. There are three important activities of plant breeding, *viz.* development of superior cultivars, seed production and maintenance breeding.
10. The significant practical achievements of plant breeding include (i) improvement in yield, (ii) improvement in quality, (iii) insect and disease resistant varieties, (iv) lodging resistant varieties, (v) early maturity, and (v) adaptability.
11. Plant breeding has several useful applications in the improvement of crop plants. However, there are some undesirable consequences of plant breeding such as reduction in biodiversity, narrow genetic base, narrow adaptation, *etc.*
12. Plant breeding has played key role in increasing the yield levels of cereals, millets and many other crops. In future, plant breeding efforts need to be directed towards the following thrust areas.
13. There is ample scope for genetic improvement of pulses and oilseed crops.
14. In India, considerable area is under rain-fed cultivation. Hence, there is need to develop crop cultivars resistant to drought conditions.
15. In view of global warming, there is need to develop crop cultivars suitable for climate change.
16. The labor has become very expensive. Hence, there is need in many crops to develop varieties suitable for machine harvesting.

17. The problem of metal toxicity is increasing in many areas particularly adjacent to industries. Hence, there is need to develop crop cultivars tolerant to metal toxicity.
18. Now new plant breeding approaches, *viz.*, molecular breeding, somatic hybridization and genetic engineering are available. In future, molecular breeding and biotechnology will supplement the conventional plant breeding methods for genetic improvement of crop plants.

C. MULTIPLE CHOICE QUESTIONS

1. Who defined plant breeding as current phase of crop evolution?

(a) J. M. Poehlman (b) N. W. Simmonds

(c) Riley (d) None of these

2. Who first made interspecific hybridization in Dianthus?

(a) Thomas Fair Child (b) N. W. Simmonds

(c) Andrew Knight (d) None of these

3. Who first made interspecific hybridization in Fruit crops?

(a) Thomas Fair Child (b) N. W. Simmonds

(c) Andrew Knight (d) None of these

4. Who first discovered sex in plants?

(a) Thomas Fair Child (b) Camerarius

(c) Andrew Knight (d) None of these

5. In which crop first transgenic plant was developed?

(a) Cotton (b) Tobacco

(c) Arabidopsis (d) None of these

6. Term Nobilization is related to which crop?

(a) Cotton (b) Tobacco

(c) Sugarcane (d) None of these

7. Term linter is related to which crop?

(a) Cotton (b) Tobacco

(c) Sugarcane (d) None of these

8. In which crop Norin 10 is a source of dwarfing gene?

(a) Wheat (b) Rice

(c) Sorghum (d) None of these

9. In which crop Dee-Geo-Woogen is a source of dwarfing gene?

(a) Wheat (b) Rice

(c) Sorghum (d) None of these

10. In which crop Kafir 60 is a source of male sterility?

(a) Wheat (b) Rice

(c) Sorghum (d) None of these

11. In which crop Tift 23A is a source of male sterility?

(a) Wheat (b) Rice

(c) Pearl millet (d) None of these

12. In which crop Gregg 399 is a source of male sterility?

(a) Cotton (b) Rice

(c) Sorghum (d) None of these

D. TRUE OR FALSE

1. Plant breeding as current phase of crop evolution by N.W. Simmonds.
2. Interspecific hybridization in Dianthus was first made by Thomas Fairchild.
3. Interspecific hybridization in Fruit crops was first made by Andrew Knight.
4. Camerarius first discovered sex in plants.
5. Transgenic plant was developed in Tobacco.
6. Term Nobilization is related to sugarcane.
7. Norin 10 is a source of dwarfing gene in wheat.
8. Dee-Geo-Woo-gen is a source of dwarfing gene in Rice.
9. Kafir 60 is a source of male sterility in Sorghum.
10. Tift 23A is a source of male sterility in pearl millet.
11. Gregg 399 is a source of male sterility in cotton.
12. Pigeon pea hybrid CPH 8 was developed from ICRISAT.

E. FILL IN THE BLANKS

1. Plant breeding is defined as current phase of crop evolution by ________.
2. Interspecific hybridization in Dianthus was first made by ________.
3. Interspecific hybridization in Fruit crops was first made by ________.
4. ________ first discovered sex in plants.
5. Transgenic plant was developed in ________.
6. Term nobilization is related to ________.

7. Norin 10 is a source of dwarfing gene in ______.
8. Dee-Geo-Woogen is a source of dwarfing gene in ______.
9. Kafir 60 is a source of male sterility in _______.
10. Tift 23A is a source of male sterility in ________.
11. Gregg 399 is a source of male sterility in _______.
12. Pigeon pea hybrid CPH 8 was developed from ________.

F. ONE WORD OR SHORT ANSWER

1. A science, an art and a technology which deals with genetic improvement of crop plants in relation to their economic use for mankind, also called as crop improvement.
2. The whole library of alleles in a crop species or sum total of genes in a species.
3. A branch of plant breeding which deals with principles and methods of improved seed production.
4. Various breeding procedures which are used for genetic improvement of crop plants in relation to their economic use.
5. Stable performance of a variety over a wide range of environmental conditions.
6. Adverse conditions imposed for growth and production of crop plants by biotic factors such as insects, diseases and parasitic weeds.
7. Adverse conditions imposed for growth and production of crop plants by environmental factors such as deficiency or excess of nutrition, moisture, temperature and light; presence of harmful gases or toxicants; and abnormal soil conditions, such as salinity, alkalinity and acidity.
8. Ability of some genotypes to give higher yield of good quality than other varieties at the same initial level of disease or insect infestation under similar environmental conditions.
9. The ratio of economic yield to biological yield. Biological yield is the total dry matter production per plant.
10. Suitability or fitness of an economic plant product in relation to its end use.
11. Cotton fabrics that are easily washed and require little ironing or pressing.

ANSWERS

C. MULTIPLE CHOICE QUESTIONS

Sl.No.	*Answer*	*Sl.No.*	*Answer*
1	(b) N. W. Simmonds	2	(a) Thomas Fair Child
3	(c) Andrew Knight	4	(b) Camerarius
5	(b) Tobacco	6	(c) Sugarcane
7	(a) Cotton	8	(a) Wheat
9	(b) Rice	10	(c) Sorghum
11	(c) Pearl millet	12	(a) Cotton

D. TRUE OR FALSE

1 to 12 all are true.

E. FILL IN THE BLANKS

Sl.No.	*Answer*	*Sl.No.*	*Answer*
1	N.W. Simmonds	2	Thomas Fairchild
3	Andrew Knight	4	Camerarius
5	Tobacco	6	Sugarcane
7	Wheat	8	Rice
9	Sorghum.	10	Pearl millet
11	Cotton	12	ICRISAT

F. ONE WORD OR SHORT ANSWER

Sl.No.	*Answer*	*Sl.No.*	*Answer*
1	Plant Breeding	2	Germplasm
3	Seed Production Technology	4	Breeding techniques
5	Adaptability	6	Biotic Stress
7	Abiotic Stress	8	Genetic Resistance
9	Harvest Index	10	Quality
11	Easy Care Property		

Chapter 2

Centers of Origin and Distribution of Crop Species

A. DEFINITIONS

Biodiversity: Total variability present within and among species of all living organisms and their habitats.

Centers of Diversity: A place, region or area where maximum variability of crop plants is observed; also called centers of origin.

Extinction: Permanent loss of a crop species due to various reasons.

Field Gene Banks: Those areas of land in which germplasm of recalcitrant crop species is maintained in the form of plants; also called plant gene banks.

Gene Banks: Various organizations where genetic diversity is maintained in living state; also called as germplasm banks.

Gene Sanctuaries: Protected areas of great genetic diversity under natural conditions *i.e. in situ* conservation.

Genetic Diversity: Total amount of genetic variation present in a population or species or variety of genes and genotypes found in a particular crop species.

Genetic Erosion: Loss of genetic diversity between and within population of a species over a period of time.

Law of Parallel Variation: This states that a particular variation observed in a crop species is also expected to be available in its another related species; also called law of homologous series of variation.

Micro Centers: Small areas within the center of diversity that exhibits tremendous genetic diversity of crop plants.

Plant genetic Resources: Genetic materials of plants which are of value as a resource for present and future generations of peoples.

Primary Centre of Diversity: The original homes of crop plants which are generally uncultivated areas such as mountains, hills, river valleys, forests, *etc.*

Secondary Centers of Diversity: The cultivated areas with vast genetic diversity of a species.

Seed gene banks: Conservation of germplasm in the form of seeds in cold

B. BACKGROUND INFORMATION

1. Genetic diversity is essential to develop improved cultivars with broad genetic base and wide adaptability. Moreover, existence of genetic diversity is also essential to meet current and future breeding requirements.
2. If a crop species has large number of genetic variants it is said to be genetically diverse. Genetic diversity provides broad genetic base to a population.
3. Genetic diversity is depleted due to genetic erosion and extinction. Therefore, collection of germplasm is essential to conserve the genetic diversity and to minimize its loss due to genetic erosion and extinction.
4. Centre of diversity refers to the geographic region in which greatest variability of a crop occurs.
5. A primary centre of diversity is the region of presumed origin, and secondary centers of diversity are regions of high diversity which have developed as a result of subsequent spread of a crop.
6. Main centers of crop diversity as proposed by Vavilov are: (1) China, (2) India (Hindustan), (3) Central Asia, (4) Asia Minor or Persia, (5) Mediterranea, (6) Abyssinia, (7) Central America or Mexico, and (8) South America.
7. There are three subsidiary centers of diversity. These are: (1) Indo- Malaya, (2) Chile, and (3) Brazil and Paraguay. All these centers are known as centers of origin or centers of diversity or Vavilovian centers of diversity.
8. The concept of parallel variation also known as law of homologous series of variation was developed by Vavilov (1951) based on his study of crop diversity and centers of origin. Law of homologous series states that a particular variation observed in a crop species is also expected to be available in its another related species.

(a) FIELD GENE BANKS IN SOME COUNTRIES

1. **Malaysia:** Oil palm has been conserved on 500 hectares.
2. **Indonesia:** Earmarked 1000 hectare area for coconut and other perennial crops.
3. **Philippines:** South East Asian germplasm of banana has been conserved.

4. **India:** Global collection of coconut has been conserved in Andman and Nicobar.

(b) GENE BANKS FOR VARIOUS CROPS IN INDIA

1. Wheat: Indian Institute of Wheat and Barley Research, Karnal.
2. Rice: Central Rice Research Institute (CRRI), Cuttack.
3. Potato: Central Potato Research Institute (CPRI), Shimla.
4. Cotton: Central Institute for Cotton Research (CICR), Nagpur
5. Pulses Kanpur Indian Institute for Pulses Research (IIPR)
6. Oilseed crops: Directorate of Oilseed Research (DOR), Hyderabad.
7. Sorghum: Directorate of Sorghum Research, Hyderabad.
8. Soybean: Directorate of Soybean Research, Indore.
9. Groundnut: Directorate of Groundnut Research, Junagarh.
10. Maize: Indian Institute of Maize Research, Pusa, New Delhi.
11. Citrus: Central Citrus Research Institute, Nagpur.
12. Sugarcane: Sugarcane Breeding Institute (SBI), Coimbatore.
13. Forage crops: Indian Gross land and Fodder Research Institute (IGFRI), Jhansi.
14. Tobacco: Central Tobacco Research Institute (CTRI), Rajamundry.
15. Tuber crops: Central Tuber Crop Research Institute (CPCRI), Trivendrum.
16. Horticultural crops: Indian Institute for Horticultural Research (IIHR), Bangalore.
17. Grapes: National Research Centre for Grapes, Pune

Table 2-1: International/Global Gene Banks for different Crops

Sl.No.	Institute	Gene Bank For
1	International Rice Research Institute (IRRI), Philippines.	Rice
2	International Wheat and Maize Improvement Centre (CIMMYT), Mexico.	Maize, Wheat, Triticale,Barley
3	International Centre for Tropical Agriculture (CIAT), Colombia.	Cassava, Beans, Rice and Maize
4	International Institute for Tropical Agriculture (IITA), Nigeria.	Cowpea, Soybean,Lima bean, Cassava, Swee tPotato
5	International Potato center, Peru.	Potato
6	International Crop Research Institute for Semiarid Tropics (ICRISAT), Hyderabad, India.	Sorghum, Pearl millet, Pigeon pea, Groundnut
7	International Centre for Agriculture Research in Dry land Areas (ICARDA), Syria.	Durum Wheat, Barley and Beans
8	Asian Vegetable Research and Development Centre (AVRDC), Taiwan.	Vegetables, Soybean,Mung bean, Sweet potato

9. Gene banks are again of two types, *viz.* (1) National gene banks, and (2) International or global gene banks. National gene banks are maintained by each country and global gene banks are located in International Crop Research Institutes/Centers. In India, gene banks are maintained by concerned Crop Research Institutes of ICAR.
10. National Bureau of Plant Genetic Resources, New Delhi is also maintaining germplasm of various field crops. Germplasm can also be conserved in the form of pollen and DNA. However, these methods are not in common use. Conservation of germplasm in the form of DNA is a difficult task. It requires sophisticated laboratory and is very expensive.

C. MULTIPLE CHOICE QUESTIONS

1. **Primary gene pool is designated as:**
   (a) GP 1 (b) GP 2
   (c) GP 3 (d) All of these
2. **Secondary gene pool is designated as:**
   (a) GP 1 (b) GP 2
   (c) GP 3 (d) All of these
3. **Tertiary gene pool is designated as:**
   (a) GP 1 (b) GP 2
   (c) GP 3 (d) All of these
4. **Centers of origin were given by::**
   (a) Darlington (b) Vavilov
   (c) Lysenko (d) None of these
5. **Law of parallel variation given by:**
   (a) Darlington (b) Vavilov
   (c) Lysenko (d) None of these
6. **Centre of origin of Maize is:**
   (a) Mexico (b) North America
   (c) West Africa (d) None of these
7. **Centre of origin of Sorghum is:**
   (a) Mexico (b) North America
   (c) Ethiopia (d) None of these
8. **Centre of origin of pearl millet is:**
   (a) Mexico (b) West Africa
   (c) South Africa (d) None of these

9. Centre of origin of potato is:

(a) South America (b) Andes region

(c) Central America (d) None of these

10. Centre of origin of Soybean is:

(a) South America (b) China

(c) Central America (d) None of these

11. Centre of origin of Tobacco is:

(a) Andes region (b) North America

(c) Central America (d) None of these

12. Centre of origin of Peanut is:

(a) Andes region (b) Brazil

(c) Central America (d) None of these

D. TRUE OR FALSE

1. Vavilov proposed eight main centers of diversity.
2. Vavilov gave three subsidiary centers of diversity.
3. Permanent loss of a crop species is called extinction.
4. Total variability present within and among species refers to biodiversity.
5. Gene pool is sum total of genes in a species.
6. Primary gene pool is designated as GP1.
7. Secondary gene pool is designated as GP2.
8. Tertiary gene pool is designated as GP3.
9. Law of parallel variation was given by Vavilov.
10. Law of parallel variation is also known as law of homologous series.

E. FILL IN THE BLANKS

1. Vavilov proposed ______ main centers of diversity.
2. Vavilov gave ______ subsidiary centers of diversity.
3. Permanent loss of a crop species is called _______.
4. Total variability present within and among species refers to _______.
5. Gene pool is ______ of genes in a species.
6. Primary gene pool is designated as _____.
7. Secondary gene pool is designated as ______.
8. Tertiary gene pool is designated as ______.

9. Law of parallel variation was given by ______.
10. Law of parallel variation is also known as law of ______.

F. ONE WORD OR SHORT ANSWER

1. Total amount of genetic variation present in a population or species or variety of genes and genotypes found in a particular crop species.
2. Permanent loss of a crop species due to various reasons.
3. Loss of genetic diversity between and within population of a species over a period of time.
4. Total variability present within and among species of all living organisms and their habitats.
5. A place, region or area where maximum variability of crop plants is observed; also called centers of origin.
6. The original homes of crop plants which are generally uncultivated areas such as mountains, hills, river valleys, forests, *etc.*
7. The cultivated areas with vast genetic diversity of a species.
8. Small areas within the center of diversity that exhibits tremendous genetic diversity of crop plants.
9. This states that a particular variation observed in a crop species is also expected to be available in its another related species; also called law of homologous series of variation.
10. Protected areas of great genetic diversity under natural conditions *i.e. in situ* conservation.
11. Various organizations where genetic diversity is maintained in living state; also called as germplasm banks.
12. Those areas of land in which germplasm of recalcitrant crop species is maintained in the form of plants; also called plant gene banks.
13. Genetic material of plants which is of value as a resource for present and future generations of people.
14. Conservation of germplasm in the form of seeds in cold

ANSWERS

C. MULTIPLE CHOICE QUESTIONS

Sl.No.	*Answer*	*Sl.No.*	*Answer*
1	(a) GP 1	2	(b) GP 2
3	(c) GP 3	4	(b) Vavilov
5	(b) Vavilov	6	(a) Mexico
7	(c) Ethiopia	8	(b) West Africa
9	(b) Andes region	10	(b) China
11	(a) Andes region	12	(b) Brazil

D. TRUE OR FALSE

1 to 10 all are true.

E. FILL IN THE BLANKS

Sl.No.	*Answer*	*Sl.No.*	*Answer*
1	Eight	2	Three
3	Extinction	4	Biodiversity
5	Sum total	6	GP1
7	GP2	8	GP3
9	Vavilov	10	Homologous series

F. ONE WORD OR SHORT ANSWER

Sl.No.	*Answer*	*Sl.No.*	*Answer*
1	Genetic Diversity	2	Extinction
3	Genetic Erosion	4	Biodiversity
5	Centers of Diversity	6	Primary Centre of Diversity
7	Secondary Centers of Diversity	8	Micro Centers
9	Law of Parallel Variation	10	Gene Sanctuaries
11	Gene Banks	12	Field Gene Banks
13	Plant genetic Resources	14	Seed gene banks

Chapter 3

Plant Genetic Resources: Conservation and Utilization

A. DEFINITIONS

Active Collections: Collections which are actively utilized in the breeding programs and are used for medium term storage (8-10 years). Seeds are stored at 0°C.

Base Collections: Total accessions available in a crop. These are used for long storage (up to 100 years). Seeds are stored at -18 or -20°C; also called principal collections.

Characterization: Recording of highly heritable phenotypic characters.

Conservation: The protection of genetic diversity from genetic erosion either under natural conditions or by storing in gene banks.

Core Collections: A set of accession derived from base collections to represent the genetic spectrum in the whole collection.

Documentation: The process of compilation, analysis, classification, storage and distribution of information.

***Ex situ* Conservation:** Preservation of germplasm in the gene banks.

Exotic Collection: The germplasm which is collected or received from other countries.

Gene Pool: The whole library of different alleles of a species or sum total of genes in a species; also called germplasm, genetic stock and genetic resources.

Genetic Erosion: Gradual reduction in genetic variability due to elimination of various genotypes. It results due to use of modern cultivars, modernization of agriculture and various development activities.

***In situ* Conservation:** Conservation of germplasm under natural conditions.

Indigenous Collection: The germplasm which is collected within the country.

Land Races. The primitive cultivars which were selected and cultivated by farmers for several generations. Land races have more genetic diversity, wider adaptability and high degree of resistance to biotic and abiotic stresses.

Modern Cultivars: The currently cultivated high yielding varieties.

Obsolete Cultivars: The improved cultivars of the recent past.

Orthodox Seeds: Seeds which can be dried to low moisture content and stored at low temperature without losing their viability.

Plant Exploration: Trips arranged for collection of germplasm from different areas.

Primary Gene Pool: The gene pool in which intermating is easy and leads to production of fertile hybrids. It includes genotypes of same species. It is designated as GP1.

Quarantine: The prophylactic measure that is used to prevent the entry of new diseases, insects and weeds from other countries.

Recalcitrant Seeds: Seeds which show very drastic loss in viability with decrease in moisture content below 12 to 13 per cent. Such species include coconut, mango, tea, coffee, rubber, jack fruit, oil palm *etc.* Such seeds cannot be conserved in seed banks.

Secondary Gene Pool: The genetic material that leads to partial fertility on crossing with primary gene pool. It includes genotypes of related species and is designated at GP2.

Tertiary Gene Pool: The genetic material which leads to production of sterile hybrids on crossing with primary gene pool. It is designated as GP3.

Working Collections: Collections which are frequently utilized by breeders in their crop improvement programs. These are stored for short term (3-5 years) at 5-10°C.

B. BACKGROUND INFORMATION

1. The sum total of genes in a crop species is referred to as genetic resources or gene pool or genetic stock or germplasm. In other words, gene pool refers to a whole library of different alleles of a species. Germplasm or gene pool is the basic material with which a plant breeder has to initiate his breeding program.
2. Gene pool represents the entire genetic variability or diversity available in a crop species.
3. Germplasm consists of land races, modern cultivars, obsolete cultivars, breeding stocks, wild forms and wild species of cultivated crops.
4. Germplasm includes both cultivated and wild species or relatives of crop plants.
5. Germplasm is collected from the centres of diversity, gene banks, gene sanctuaries, farmers' fields, markets and seed companies.

6. Germplasm is the basic material for launching a crop improvement program.
7. Germplasm may be indigenous (collected within country) or exotic (collected from foreign countries).
8. The germplasm consists of various plant materials of a crop such as (1) land races, (2) obsolete cultivars, (3) modern cultivars, (4) advanced (homozygous) breeding materials, (5) wild forms (races) of cultivated species, (6) wild relatives, and (7) mutants.
9. Those naturally occurring plant species which have common ancestry with crops and can cross with crop species are referred to as wild relatives or wild species.
10. Wild relatives are important sources of resistance to biotic (diseases and insects) and abiotic (drought, cold, frost, salinity, *etc.*) stresses.
11. However, wild relatives are used as the last resort in crop improvement programs, because their use in crossing leads to : (1) hybrid sterility, (2) hybrid in-viability and (3) transfer of several undesirable genes to the cultivated species along with desirable alleles. This group constitutes a minor part of gene pool. Interspecific derivatives are added to the gene pool.
12. Based on the use and duration of conservation, seed collections are of three types: (1) base collections, (2) active collections, and (3) working collections.
13. There are six important activities related to plant genetic resources: *viz.* (1) exploration and collection, (2) conservation (3) evaluation, (4) documentation, (5) distribution, and (6) utilization.
14. There are five important sources of germplasm collection : *viz.* (1) centres of diversity, (2) gene banks, (3) gene sanctuaries, (4) seed companies, and (5) farmers' fields. Moreover, collections can be made by local exploration trips to the regions of crop diversity.
15. Seeds which can be dried to low moisture content and stored at low temperature without losing their viability are known as orthodox seeds. This group includes seeds of corn, wheat, rice, carrot, beets, papaya, pepper, chickpea, lentil, soybean, cotton, sunflower, various beans, eggplant and all the *brassicas.* These seeds can be dried and stored at low temperatures for long periods of time.
16. Seeds which show very drastic loss in viability with a decrease in moisture content below 12 to 13 per cent are known as recalcitrant seeds. This group includes cocoa, coconut, mango, tea, coffee, rubber, jack fruit and oil palm seeds.
17. There are two types of organizations, *viz.* International and National which are associated with germplasm. International Plant Genetic Resources Institute (IPGRI), Rome, Italy coordinates at global level. Various

International Crop Research Institutes deal with germplasm of concerned crops.

18. In India, National Bureau of Plant Genetic Resources (NBPGR). New Delhi deals with various aspects of germplasm of agricultural and horticultural crops. Forest Research Institute, Dehradun deals with germplasm of forest species and Botanical Survey of India, Kolkata deals with germplasm of remaining plant species.
19. There are chances of the entry of new diseases, insects and weeds when the material is collected and introduced from other countries. Several new weeds, diseases and insects have entered in India in the past along-with plant collections. To prevent entry of new diseases, insects and weeds an Act was formulated in India in 1914 which is referred to as Destructive Insects and Pest Act.

C. MULTIPLE CHOICE QUESTIONS

1. Which of the following is called primary gene pool?

(a) GP 1 (b) GP 2
(c) GP 3 (d) All of these

2. Which of the following is called secondary gene pool?

(a) GP 1 (b) GP 2
(c) GP 3 (d) All of these

3. Which of the following is called tertiary gene pool?

(a) GP 1 (b) GP 2
(c) GP 3 (d) All of these

4. Which of the following is recalcitrant seed?

(a) Cocoa (b) Coconut
(c) Mango (d) All of these

5. Which of the following is recalcitrant seed?

(a) Tea (b) Coffee
(c) Rubber (d) All of these

6. Which of the following is an orthodox seed?

(a) Wheat (b) Barley
(c) Rice (d) All of these

7. Which of the following is recalcitrant seed?

(a) Chick pea (b) Pigeon pea
(c) Coconut (d) All of these

8. Recording observations on highly heritable traits refers to:

(a) Screening (b) Characterization

(c) Evaluation (d) All of these

9. With plant introduction, there are chances of entry of:

(a) New disease (b) New insect

(c) New Weed (d) All of these

10. Plant quarantine prevents entry of:

(a) New disease (b) New insect

(c) New Weed (d) All of these

11. Which of the following is responsible for introduction of agricultural plants?

(a) NBPGR, New Delhi (b) FRI, Dehradun

(c) BSI, Kolkata (d) All of these

12. Which of the following is responsible for introduction of Forest plants?

(a) NBPGR, New Delhi (b) FRI, Dehradun

(c) BSI, Kolkata (d) All of these

13. Which of the following is responsible for introduction of botanical and medicinal plants?

(a) NBPGR, New Delhi (b) FRI, Dehradun

(c) BSI, Kolkata (d) All of these

14. Which of the following is responsible for introduction of horticultural plants?

(a) NBPGR, New Delhi (b) FRI, Dehradun

(c) BSI, Kolkata (d) All of these

D. TRUE OR FALSE

1. Land races are primitive cultivars.
2. Improved varieties of recent past are known as obsolete cultivars,
3. The currently cultivated high yielding varieties are referred to as modern cultivars.
4. Primary gene pool is designated as GP1.
5. Secondary gene pool is designated as GP2.
6. Tertiary gene pool is designated as GP3.
7. Conservation of germplasm under natural conditions is called *in-situ* conservation.

8. Preservation of germplasm in the gene banks is called ex-situ conservation.
9. Total accessions available in a crop refer to base collections.
10. Recording of observations on highly heritable phenotypic characters is called characterization.
11. NBPGR is located at New Delhi.
12. FRI is located at Dehradun.
13. BSI is located at Kolkata.
14. IPGRI is located in Rome, Italy.

E. FILL IN THE BLANKS

1. Land races are _______ cultivars.
2. Improved varieties of recent past are known as ________ cultivars,
3. The currently cultivated high yielding varieties are referred to as _______ cultivars.
4. Primary gene pool is designated as ________)_.
5. Secondary gene pool is designated as _______.
6. Tertiary gene pool is designated as ________.
7. Conservation of germplasm under natural conditions is called _________.
8. Preservation of germplasm in the gene banks is called ________.
9. Total accessions available in a crop refer to _____collections.
10. Recording of observations on highly heritable phenotypic characters is called ______.
11. NBPGR is located at _______.
12. FRI is located at ________.
13. BSI is located at ____.
14. IPGRI is located in _______.

F. ONE WORD OR SHORT ANSWER

1. The whole library of different alleles of a species or sum total of genes in a species; also called germplasm, genetic stock and genetic resources.
2. The germplasm which is collected within the country.
3. The germplasm which is collected or received from other countries.
4. Total accessions available in a crop. These are used for long storage (up to 100 years). Seeds are stored at -18 or -20°C; also called principal collections.
5. Collections which are actively utilized in the breeding programs and are used for medium term storage (8-10 years). Seeds are stored at 0°C.

6. Collections which are frequently utilized by breeders in their crop improvement programs. These are stored for short term (3-5 years) at 5-10°C.
7. A set of accession derived from base collections to represent the genetic spectrum in the whole collection.
8. Recording of highly heritable phenotypic characters.
9. The gene pool in which inter-mating is easy and leads to production of fertile hybrids. It includes genotypes of same species. It is designated as GP1.
10. The genetic material that leads to partial fertility on crossing with primary gene pool. It includes genotypes of related species and is designated at GP2.
11. The genetic material which leads to production of sterile hybrids on crossing with primary gene pool. It is designated as GP3.
12. The primitive cultivars which were selected and cultivated by farmers for many generations. Land races have more genetic diversity, wider adaptability and high degree of resistance to biotic and abiotic stresses.
13. The improved cultivars of the recent past.
14. The currently cultivated high yielding varieties.
15. Trips arranged for collection of germplasm from different areas.
16. Gradual reduction in genetic variability due to elimination of various genotypes. It results due to use of modern cultivars, modernization of agriculture and various development activities.
17. The protection of genetic diversity from genetic erosion either under natural conditions or by storing in gene banks.
18. Conservation of germplasm under natural conditions.
19. Preservation of germplasm in the gene banks.
20. Seeds which can be dried to low moisture content and stored at low temperature without losing their viability.
21. Seeds which show very drastic loss in viability with decrease in moisture content below 12 to 13 per cent. Such species include coconut, mango, tea, coffee, rubber, jack fruit, oil palm *etc.* Such seeds cannot be conserved in seed banks.
22. The process of compilation, analysis, classification, storage and distribution of information.
23. The prophylactic measure that is used to prevent the entry of new diseases, insects and weeds from other countries.

ANSWERS

C. MULTIPLE CHOICE QUESTIONS

Sl.No.	*Answer*	*Sl.No.*	*Answer*
1	(a) GP 1	2	(b) GP 2
3	(c) GP 3	4	(d) All of these
5	(d) All of these	6	(d) All of these
7	(c) Coconut	8	(b) Characterization
9	(d) All of these	10	(d) All of these
11	(a) NBPGR, New Delhi	12	(b) FRI, Dehradun
13	(c) BSI, Kolkata	14	(a) NBPGR, New Delhi

D. TRUE OR FALSE

1 to 14 all are true.

E. FILL IN THE BLANKS

Sl.No.	*Answer*	*Sl.No.*	*Answer*
1	Primitive	2	Obsolete
3	Modern	4	GP1
5	GP2	6	GP3
7	In-situ conservation	8	Ex-situ conservation
9	Base	10	Characterization
11	New Delhi	12	Dehradun
13	Kolkata	14	Rome, Italy

F. ONE WORD OR SHORT ANSWER

Sl.No.	*Answer*	*Sl.No.*	*Answer*
1	Gene Pool	2	Indigenous Collection
3	Exotic Collection	4	Base Collections
5	Active Collections	6	Working Collections
7	Core Collections	8	Characterization
9	Primary Gene Pool	10	Secondary Gene Pool
11	Tertiary Gene Pool	12	Land Races
13	Obsolete Cultivars	14	Modern Cultivars
15	Plant Exploration	16	Genetic Erosion
17	Conservation	18	In situ Conservation
19	Ex situ Conservation	20	Orthodox Seeds
21	Recalcitrant Seeds	22	Documentation
23	Quarantine		

Chapter 4

Genetics of Qualitative and Quantitative Characters

A. DEFINITIONS

Additive variance: Average effect of genes on all segregating loci. Such genes show lack of dominance, *i.e.,* intermediate expression.

Additive x Additive Epistasis: Interaction between two or more loci each exhibiting lack of dominance individually. It is denoted as *AA* and is fixable.

Additive x Dominance Epistasis: Interaction between two or more loci, one exhibiting lack of dominance and the other dominance. It is denoted as *AD* and is non-fixable.

Character: Any property of an individual showing heritable variation. It includes morphological, physiological, biochemical and behavioral properties.

Dominance Variance: Deviation from the mean value due to intra-allelic interaction.

Dominance x Dominance Epistasis: Interaction of two or more loci, each exhibiting dominance individually. It is represented as *DD* and is non-fixable.

Environmental Variation: The non-heritable variation which is entirely due to environmental effects and varies under different environmental conditions.

Epistatic Variance: Deviation from the mean value as a consequence of non-allelic interaction, *i.e.,* interaction between two or more genes. It is of three types *viz.,* additive x additive, additive x dominance and dominance x dominance.

Genetic Variance: Heritable portion of total or phenotypic variance. It is of three types, *viz.* additive, dominance and epistatic variances.

Genotypic Variation: The inherent or genetic variation which remains unaltered by environmental changes.

Oligogenic Trait: Characters which are controlled by one or few genes each having detectable individual effect and exhibit discontinuous variation.

Phenotypic Variation: The total variability which is observable.

Polygenic Traits: Characters which are governed by several genes each having small individual effect and exhibit discontinuous variation.

B. BACKGROUND INFORMATION

1. In genetics, variation is of three types, *viz.*, phenotypic, genotypic and environmental. Phenotypic variation includes both genotypic and environmental components and hence changes under different environmental conditions.
2. Phenotypic variation is measured in terms of phenotypic variance.
3. Environmental variation refers to non-heritable variation which is entirely due to environmental effects and varies under different environmental conditions. This is uncontrolled variation which is measured in terms of error mean variance.
4. Environmental variation in true breeding parental lines and their F_1 hybrid is non-heritable.
5. R.A. Fisher [1918] divided the genetic variance in to three components, *viz.*, additive, dominance and epistatic variances.
6. Wright [1935] classified the genetic variance in to two fractions, *viz.*, additive variance and non-additive variance [dominance and epistatic variances].
7. Mather [1949] classified the genetic variance in to two components, *viz.*, heritable fixable [additive variance and additive x additive type of epistasis] and heritable non-fixable [dominance variance and AD and DD types of epistasis].
8. Additive genes show lack of dominance, *i.e.* intermediate expression. It is a measure of additive gene action.
9. In natural plant breeding population, additive genetic variance is pre-dominant one closely followed by dominance variance.
10. Additive genetic variance is associated with homozygosity and, therefore, it is expected to be maximum in self-pollinating species and minimum in cross pollinating species.
11. It is measured by various biometrical techniques. The general combining ability [gca] is a measure of additive gene effects.
12. Selection is very effective with additive genetic variance, because it is fixable. The additive variance gets depleted proportionate to the improvement made by selection.
13. Additive gene action is a measure of the breeding value of a genotype. It is used for estimation of heritability in narrow sense.

14. Transgressive segregation results due to additive gene action when the parents are intermediate to extreme values of the segregating population.
15. Dominant genes show incomplete dominance, complete dominance or over-dominance. The heterozygote approaches more closely to one of the parents for a character.
16. In natural plant breeding population, dominance genetic variance is always lesser than additive variance.
17. Dominance genetic variance is associated with heterozygosity and, therefore, it is expected to be maximum in cross pollinating species and minimum in self- pollinating species.
18. It is measured by various biometrical techniques. The specific combining ability [sca] is a measure of dominance variance.
19. Dominance variance is non-fixable and ,therefore, selection for traits governed by dominance variance is not effective.
20. Dominance variance is the chief cause of heterosis or hybrid vigor.
21. Dominance variance gets depleted though selfing or inbreeding.
22. Epistatic genes interact and sometimes lead to new phenotypic expression. The genes may be dominant or recessive.
23. In natural plant breeding population, epistatic genetic variance has the lowest magnitude.
24. Epistatic variance includes both additive and non-additive components.
25. Epistatic variance is of three types, *viz.*, additive x additive, additive x dominance and dominance x dominance.
26. The additive x additive type of epistasis is fixable. Hence, selection is effective for traits governed by this type of epistasis.
27. Heterosis breeding would be rewarding with additive x dominance and dominance x dominance types of epistasis, because these are non-fixable.
28. Epistatic variance can be measured by various biometrical techniques such as generation mean analysis, triallel analysis, quadriallel analysis, *etc.*

C. MULTIPLE CHOICE QUESTIONS

1. Qualitative traits are also known as:

(a) Oligogenic traits (b) Monogenic traits
(c) Major gene traits (d) All of these

2. Qualitative traits exhibit:

(a) Discrete variation (b) Highly heritability
(c) No transgression (d) All of these

3. **Quantitative traits are also known as:**
   (a) Polygenic traits (b) Low heritable traits
   (c) Minor gene traits (d) All of these

4. **Quantitative traits exhibit:**
   (a) Continuous variation (b) Low heritability
   (c) Transgression (d) All of these

5. **Which of the following genetic variance is fixable?**
   (a) Additive (b) Dominance
   (c) Epistatic (d) All of these

6. **Selection is effective with which of the following genetic variance?**
   (a) Additive (b) Dominance
   (c) Epistatic (d) All of these

7. **Which of the following epistatic variance is fixable?**
   (a) Additive x additive (b) Dominance x dominance
   (c) Additive x dominance (d) All of these

8. **Which of the following epistatic variance responds to selection?**
   (a) Additive x additive (b) Dominance x dominance
   (c) Additive x dominance (d) All of these

9. **Dominant genes may show:**
   (a) Incomplete dominance (b) Complete dominance
   (c) over dominance (d) All of these

10. **Which of the following gene action is associated with transgressive segregation?**
    (a) Additive (b) Dominance
    (c) Epistatic (d) All of these

11. **Which of the following genetic variance has the highest magnitude in natural plant breeding population?**
    (a) Additive (b) Dominance
    (c) Epistatic (d) All of these

12. **Which of the following genetic variance has the lowest magnitude in natural plant breeding population?**
    (a) Additive (b) Dominance
    (c) Epistatic (d) All of these

13. Which of the following genetic variance has intermediate magnitude in natural plant breeding population?

(a) Additive (b) Dominance

(c) Epistatic (d) All of these

14. Which of the following genetic variance is associated with heterosis?

(a) Additive (b) Dominance

(c) Epistatic (d) All of these

D. TRUE OR FALSE

1. Dominance variance is the chief cause of heterosis.
2. Dominance variance gets depleted though inbreeding.
3. The additive x additive type of epistasis is fixable.
4. Dominance variance is non-fixable.
5. Transgressive segregation is associated with additive gene action.
6. In natural plant breeding population, epistatic genetic variance has the lowest magnitude.
7. In natural plant breeding population, additive genetic variance has the highest magnitude.
8. Additive gene action is a measure of the breeding value of a genotype.
9. Wright [1935] classified the genetic variance in to two fractions, *viz.*, additive variance and non-additive variance.
10. Mather [1949] classified the genetic variance in to two components, *viz.*, heritable fixable and heritable non-fixable.

E. FILL IN THE BLANKS

1. Dominance variance is the chief cause of _______.
2. _____ variance gets depleted though inbreeding.
3. The additive x additive type of epistasis is ______.
4. Dominance variance is _______.
5. Transgressive segregation is associated with ________ action.
6. In natural plant breeding population, epistatic genetic variance has the _____ magnitude.
7. In natural plant breeding population, additive genetic variance has the _____magnitude.
8. Additive gene action is a measure of the breeding value of a _______.
9. ________classified the genetic variance in to two fractions, *viz.*, additive variance and non-additive variance.

10. ______ classified the genetic variance in to two components, *viz.*, heritable fixable and heritable non-fixable.

F. ONE WORD OR SHORT ANSWER

1. Any property of an individual showing heritable variation. It includes morphological, physiological, biochemical and behavioral properties.
2. Characters which are controlled by one or few genes each having detectable individual effect and exhibit discontinuous variation.
3. Characters which are governed by several genes each having small individual effect and exhibit discontinuous variation.
4. The total variability which is observable.
5. The inherent or genetic variation which remains unaltered by environmental changes.
6. The non-heritable variation which is entirely due to environmental effects and varies under different environmental conditions.
7. Heritable portion of total or phenotypic variance. It is of three types, *viz.* additive, dominance and epistatic variances.
8. Average effect of genes on all segregating loci. Such genes show lack of dominance, *i.e.*, intermediate expression.
9. Deviation from the mean value due to intra-allelic interaction.
10. Deviation from the mean value as a consequence of non-allelic interaction, *i.e.*, interaction between two or more genes. It is of three types *viz.*, additive x additive, additive x dominance and dominance x dominance.
11. Interaction between two or more loci each exhibiting lack of dominance individually. It is denoted as *AA* and is fixable.
12. Interaction between two or more loci, one exhibiting lack of dominance and the other dominance. It is denoted as *AD* and is non-fixable.
13. Interaction of two or more loci, each exhibiting dominance individually. It is represented as *DD* and is non-fixable.

ANSWERS

C. MULTIPLE CHOICE QUESTIONS

Sl.No.	Answer	Sl.No.	Answer
1	(d) All of these	2	(d) All of these
3	(d) All of these	4	(d) All of these
5	(a) Additive	6	(a) Additive
7	(a) Additive x additive	8	(a) Additive x additive
9	(d) All of these	10	(a) Additive
11	(a) Additive	12	(c) Epistatic
13	(b) Dominance	14	(b) Dominance

D. TRUE OR FALSE

1 to 10 all are true.

E. FILL IN THE BLANKS

Sl.No.	Answer	Sl.No.	Answer
1	Heterosis	2	Inbreeding
3	Fixable	4	Non-fixable
5	Additive gene	6	Lowest
7	Highest	8	Genotype
9	Wright	10	Mather

F. ONE WORD OR SHORT ANSWER

Sl.No.	Answer	Sl.No.	Answer
1	Character	2	Oligogenic Trait
3	Polygenic Traits	4	Phenotypic Variation
5	Genotypic Variation	6	Environmental Variation
7	Genetic Variance	8	Additive variance
9	Dominance Variance	10	Epistatic Variance
11	Additive × Additive Epistasis	12	Additive × Dominance Epistasis
13	Dominance × Dominance Epistasis		

Section II

Breeding Methods

Chapter 5

Genetic Basis of Breeding Methods

A. DEFINITIONS

Allogamy: Development of seed by cross pollination.

Autogamy: Development of seed by self-pollination.

Heterogeneous Population: A population that is composed of genetically dissimilar plants such as land races, mass selected populations, composites, synthetics and multi-lines.

Heterozygous Population: A Population that segregates on selfing such as *F*1 hybrids, composites and synthetics. Heterozygous individuals have unlike alleles at the corresponding loci.

Homogeneous Population: A population of genetically similar plants such as pure line, *F*1 between two pure lines and progeny of a clone.

Homozygosity: The proportion of homozygous individuals in a segregating population. Homozygosity = $[(2m-1)/2m]n$, where m = number of generations of self- pollination and n = number of gene pairs segregating.

Homozygous Population: A true breeding population such as pure lines and mass selected populations in self-pollinated species. Homozygous individuals have identical alleles at the corresponding loci.

Plant Breeding Methods: Various procedures (Selection, hybridization, mutation, *etc.*) that are used for genetic improvement of crop plants.

B. BACKGROUND INFORMATION

1. Based on mode of pollination, crop plants are divided into two groups, *viz.* (1) self-pollinated, and (2) cross pollinated.
2. Based on mode of reproduction crop plants are of two types, *viz.* (1) seed propagated, and (2) vegetatively propagated.

3. Thus, crop plants are divided into three groups, *viz.*(1) self-pollinated, (2) cross pollinated, and (3) vegetatively propagated.
4. Self-pollinated species are also known as autogamous species or inbreeders. Various plant characters such as homogamy, cleistogamy, chasmogamy, bisexuality *etc.* favor self- fertilization.
5. Cross-pollinated species are also referred to as allogamous species or outbreeders. Various plant characters which promote cross pollination include dichogamy, moecy, dioecy, heterostyly, herkogamy, self-incompatibility and male sterility.
6. Some crop plants propagate by asexual means *i.e.* by stem or root cuttings or by other means. Such species are known as asexually propagated species or vegetatively propagated species. Such species are found in both self and cross pollinated groups. Generally, asexually propagated species are highly heterozygous and have broad genetic base, wide adaptability and more flexibility.
7. Based on genetic constitution, plant breeding populations are of four types, *viz.* (1) homogeneous, (2) heterogeneous, (3) homozygous, and (4) heterozygous.
8. Examples of homogeneous populations are pure lines, inbred lines, F_1 hybrid between two pure lines or inbred lines and progeny of a clone.
9. Examples of heterogeneous populations are land races, mass selected populations, composites, synthetics and multi-lines.
10. Examples of homozygous homogeneous populations are pure lines, and inbred lines.
11. Examples of homozygous and heterogeneous populations are mass selected populations multiline in self-pollinated plants.
12. General breeding methods include Plant Introduction, Pure line selection, mass selection, progeny selection, pedigree method, bulk method, back cross method, SSD, clonal selection, heterosis breeding, synthetics and composites.
13. Special breeding methods include: mutation breeding, polyploidy breeding, wide crossing or distant hybridization, transgenic breeding, molecular breeding and somatic hybridization.
14. Those breeding approaches which are used for accumulating favorable gene in a population are called population improvement approaches. Such approaches include: recurrent selection, disruptive selection, diallel selective mating system and biparental mating.
15. Methods Involving hybridization include, Pedigree, bulk, backcross and SSD Methods; heterosis breeding and development of synthetics, composites. And population improvement approaches.

16. Methods not Involving hybridization include plant Introduction, pure line selection, mass selection, progeny selection, clonal selection, mutation breeding and transgenic breeding.
17. Plant breeding methods that are used for genetic improvement of self-pollinated or autogamous species include: (1) Plant Introduction, (2) Pure line selection, (3) Mass selection, (4) Pedigree method, (5) Bulk method, (6) Single seed descent method, (7) Backcross method, (8) Heterosis breeding, (9) Mutation breeding, (10) Polyploidy breeding, (11) Distant hybridization and (12) Transgenic breeding.
18. Breeding methods that are used for genetic improvement of cross pollinated or allogamous species include: (1) Plant Introduction, (2) Mass and progeny selection, (3) Backcross method, (4) Heterosis breeding, (5) Synthetic breeding, (6) Composite breeding, (7) Polyploidy breeding, (8) Distant hybridization, and (9) Transgenic breeding.
19. Important breeding methods applicable to asexually propagated species are: (1) Plant Introduction, (2) clonal selection, (3) Mass selection, (4) Heteross breeding, (5) Mutation breeding, (6) Polyploidy breeding, (7) Distant hybridization, and (8) Transgenic breeding.
20. Thus a pure line variety is homozygous and homogeneous population.
21. In self-pollinated crops, a mass selected variety is a mixture of several pure lines. Thus it is a homozygous but heterogeneous population.
22. Progeny selection is used in cross pollinated species. A variety developed by this method is heterozygous and heterogeneous population because it consists of several hetero and homozygotes.
23. In self-pollinated crops progeny of a single best homozygote is released as a variety. Thus a variety developed by pedigree method has a homozygous and homogeneous population.
24. Varieties developed by Bulk and SSD methods are homozygous and homogeneous.
25. In self-pollinated species, varieties developed by back cross method are homozygous and homogeneous.
26. A multiline variety is a homozygous but heterogeneous population.
27. A clonal variety has heterozygous but homogeneous population.
28. A hybrid variety has homogeneous but heterozygous population.
29. Synthetic and composite varieties consist of several homozygotes and heterozygotes and thus constitute a heterogeneous population.
30. Transgenic variety is similar to parent variety except for the gene incorporated.

C. MULTIPLE CHOICE QUESTIONS

1. **Special breeding methods include:**
   (a) Mutation breeding (b) Polyploidy breeding
   (c) Distant hybridization (d) All of these

2. **Special breeding methods include:**
   (a) Transgenic breeding (b) Molecular breeding
   (c) Somatic hybridization (d) All of these

3. **Population improvement approaches include:**
   (a) Recurrent selection (b) Disruptive selection
   (c) Diallel selective mating (d) All of these

4. **Breeding methods involving hybridization include:**
   (a) Pedigree method (b) Bulk method
   (c) Back cross method (d) All of these

5. **Breeding methods is not involving hybridization include:**
   (a) Pure line selection (b) Mass selection
   (c) Progeny (d) All of these

6. **Breeding methods is not involving hybridization include:**
   (a) Mutation breeding (b) Transgenic breeding
   (c) Molecular breeding (d) All of these

7. **Homogeneous populations include:**
   (a) Pure lines (b) Inbred lines
   (c) F_1 hybrids (d) All of these

8. **Heterogeneous populations include:**
   (a) Land Races (b) Composite varieties
   (c) Synthetic varieties (d) All of these

9. **Heterozygous populations include:**
   (a) F_1 hybrids (b) Composite varieties
   (c) Synthetic varieties (d) All of these

10. **Which of the following is homogeneous population?**
    (a) Land Races (b) Pure lines
    (c) Synthetic varieties (d) All of these

D. TRUE OR FALSE

1. In self-pollinated crops progeny of a single best homozygote is released as a variety.
2. A variety developed by pedigree method has a homozygous and homogeneous population.
3. Varieties developed by Bulk and SSD methods are homozygous and homogeneous.
4. In self-pollinated species, varieties developed by back cross method are homozygous and homogeneous.
5. A multiline variety is a homozygous but heterogeneous population.
6. A clonal variety has heterozygous but homogeneous population.
7. A hybrid variety has homogeneous but heterozygous population.
8. Synthetic and composite varieties consist of several homozygotes and heterozygotes.
9. Synthetic and composite varieties constitute a heterogeneous population.
10. Transgenic variety is similar to parent variety except for the gene incorporated.

E. FILL IN THE BLANKS

1. In self-pollinated crops progeny of a single best ______ is released as a variety.
2. A variety developed by pedigree method has a homozygous and ______ population.
3. Varieties developed by Bulk and SSD methods are homozygous and ______.
4. In self-pollinated species, varieties developed by back cross method are homozygous and ________.
5. A multiline variety is a homozygous but ________ population.
6. A clonal variety is heterozygous but _______population.
7. A hybrid variety has homogeneous but _______ population.
8. Synthetic and composite varieties consist of several _________.
9. Synthetic and composite varieties constitute a _______ population.
10. Transgenic variety is similar to parent variety except for the gene _______.

F. ONE WORD OR SHORT ANSWER

1. Various procedures (Selection, hybridization, mutation, *etc.*) that are used for genetic improvement of crop plants.
2. A population of genetically similar plants such as pure line, *F*1 between two pure lines and progeny of a clone.

3. The proportion of homozygous individuals in a segregating population. Homozygosity = $[(2m - 1)/2m]n$, where m = number of generations of self-pollination and n = number of gene pairs segregating.
4. A population that is composed of genetically dissimilar plants such as land races, mass selected populations, composites, synthetics and multi-lines.
5. A true breeding population such as pure lines and mass selected populations in self-pollinated species. Homozygous individuals have identical alleles at the corresponding loci.
6. A Population that segregates on selfing such as $F1$ hybrids, composites and synthetics. Heterozygous individuals have unlike alleles at the corresponding loci.
7. Development of seed by self-pollination.
8. Development of seed by cross pollination.

ANSWERS

C. MULTIPLE CHOICE QUESTIONS

Sl.No.	Answer	Sl.No.	Answer
1	(d) All of these	2	(d) All of these
3	(d) All of these	4	(d) All of these
5	(d) All of these	6	(d) All of these
7	(d) All of these	8	(d) All of these
9	(d) All of these	10	(b) Pure lines

D. TRUE OR FALSE

1 to 10 all are true.

E. FILL IN THE BLANKS

Sl.No.	Answer	Sl.No.	Answer
1	Homozygote	2	Homogeneous
3	Homogeneous	4	Homogeneous
5	Heterogeneous	6	Homogeneous
7	Heterozygous	8	Heterozygotes
9	Heterogeneous	10	Incorporated

F. ONE WORD OR SHORT ANSWER

Sl.No.	Answer	Sl.No.	Answer
1	Plant Breeding Methods	2	Homogeneous Population
3	Homozygosity	4	Heterogeneous Population
5	Homozygous Population	6	Heterozygous Population
7	Autogamy	8	Allogamy

Chapter 6

Breeding Self-pollinated Crops (Plant Introduction and Pure Line Selection)

A. DEFINITIONS

Acclimatization: Adaptation of an introduced variety to the new environment.

Artificial Selection: The selection made by human. It favors those characteristics of plants that are related to yield and quality.

Characterization: Recording observations on highly heritable traits such as color, shape, hairiness *etc.*

Cyclic Selection: Selection in one direction for one generation or season and in opposite direction in next generation or season.

Direct Introduction: Introductions which are immediately adapted to the changed environment.

Directional Selection: Selection in favor of an extreme phenotype such as earliness or lateness, dwarfness or tallness.

Domestication. The process of bringing wild and weedy species under human management.

Exotic Variety: A foreign variety which is directly recommended for commercial cultivation.

Indirect Introduction: Introductions which take some years for adaptation in changed environment.

Natural Selection: Selection that operates in nature without human interference. It favors those plant characters that are essential for survival (adaptation) of a species.

Plant Introduction: Transposition of crop plants from the place of their cultivation to such areas where they were never grown earlier.

Primary introduction: Introduction that can be used for commercial cultivation without any change in the original genotype.

Pure line Selection: Development of new variety through identification and isolation of a single best plant progeny.

Pure Line: Progeny of a self-pollinated homozygous plant obtained by selfing.

Secondary Introduction: Introduction that can be used as a variety after selection from original genotype or used for transfer of some desirable gene to the cultivated variety.

Selection: The process that favors survival and further propagation of some plants having more desirable characters than others. It is of two types, *viz.*, natural and artificial.

Stabilizing Selection: Selection in favor of intermediate types; also called centripetal selection.

B. BACKGROUND INFORMATION

(i) Plant Introduction

1. Plant introduction is an oldest method of crop improvement. It is used in all three groups of crops. *viz.* self-pollinated, cross pollinated and asexually propagated. In other words, this method is applicable to all three types of crop species.
2. Earlier, the introduction of plants from one place to another used to be done by travellers, traders, invaders and merchants. Now specific organization have been established for this work (see later).Now the plant material is obtained through expeditions, personal visit, correspondence, exchange of material, purchase or as a gift.
3. The material of seed propagated crops is introduced in the form of seed and that of vegetatively propagated crops in the form of cuttings or propagules.
4. There are five important sources of plant collections : *viz.* (1) centres of diversity, (2) gene banks, (3) gene sanctuaries, (4) seed companies, and (5) farmers' fields.
5. Based on adaptation, introductions are of two types, *viz.* (1) primary introductions, and (2) secondary introductions. Based on utilization, again introduction are of two types, *viz.* (*i*) direct introduction, and (*ii*) Indirect Introduction.
6. In wheat, varieties Sonora 64, Lerma Rojo are examples of direct release of introduced material in India. These varieties were introduced from Mexico and released directly for commercial cultivation in India.

7. In rice, semi-dwarf rice, varieties Taichung Native 1, IR 8, IR 20 and IR 36 are examples of primary Introduction. Many such examples can be cited from other field crops and fruit crops.
8. In wheat, varieties Kalyan Sona and Sonalika were released after selection from the material received from Mexico.
9. Plant material is introduced for five main purposes, *viz.* (1) economic use, (2) study of origin and evolution of crop plants, (3) conservation of diversity, (4) genetic improvement of crop plants, and (5) aesthetic interest.
10. IPGRI is located at Rome, Italy which deals with Introduction of plant material on global level. Various International crop research Institutes also undertake Plant Introduction work in collaboration with IPGRI.
11. In India, plant Introduction work is undertaken by National Bureau of Plant Genetic Resources (NBPGR), New Delhi. NBPGR which deals with the introduction of Agricultural and Horticultural crops. In India, two more organizations, *viz.* (1) Forest Research Institute (FRI), Dehradun, and (2) Botanical Survey of India (BSI), Kolkata are also concerned with plant introduction work.
12. Plant introductions are utilized in crop improvement in three main ways : *viz.* (1) direct as a variety, (2) as a variety after selection, and (3) as a parent in the hybridization for the development of improved variety.
13. The main demerit of this method is that there are chances of the entry of new diseases, insects and weeds in the country along with the introduced material. For example, *Argimone mexicana* weed, late blight of potato, bunchy top of banana, and coffee rust disease, and woolly aphis of apple and potato tuber moth insects were introduced in India along with introduced material. However, all these entered before the establishment of quarantine organization.

(ii) Pure Line Selectiion

14. The concept of pure line theory was developed by Johannsen (1903, 1926), a Danish biologist, working with Princess Variety of common bean (*Phaseolus vulgaris*).
15. Pure line selection consists of four major steps, *viz.* (1) selection of a heterogeneous population from which pure lines have to be isolated, (2) isolation of pure lines by individual plant selection, (3) testing of pure lines in field trials, and (4) release of the best pure line as a variety.
16. Pure line selection is a good method of isolating the best genotype for yield, disease resistance, insect resistance, earliness, quality, *etc.* from a heterogeneous, or mixed population of an old variety.
17. The varieties developed by pure line selection have poor adaptability due to narrow genetic base. All the plants of a pure line have identical genotypes. Hence, such varieties are more prone to the attack of new disease due to genetic uniformity.

18. Superior varieties have been isolated by pure line selection from the heterogeneous populations in several self-pollinated crops like wheat, barley, paddy, peanut, chickpea, black gram, green gram, linseed, cotton and tobacco.
19. Pure lines are used either as varieties or as parents in hybridization for development of superior varieties or hybrids in self-pollinated species.

C. MULTIPLE CHOICE QUESTIONS

1. Which organization is responsible for plant introduction in India?

(a) NBPGR, New Delhi (b) FRI, Dehradun
(c) BSI, Kolkata (d) All of these

2. Which organization is responsible for introduction of agricultural and horticultural crops?

(a) NBPGR, New Delhi (b) FRI, Dehradun
(c) BSI, Kolkata (d) All of these

3. Which organization is responsible for introduction of forest trees?

(a) NBPGR, New Delhi (b) FRI, Dehradun
(c) BSI, Kolkata (d) All of these

4. Which organization is responsible for introduction of medicinal and botanical plants?

(a) NBPGR, New Delhi (b) FRI, Dehradun
(c) BSI, Kolkata (d) All of these

5. National Bureau of Plant Genetic Resources is located in:

(a) New Delhi (b) Dehradun
(c) Kolkata (d) none of these

6. Forest Research Institute is located in:

(a) New Delhi (b) Dehradun
(c) Kolkata (d) none of these

7. Botanical Survey of India is located in:

(a) New Delhi (b) Dehradun
(c) Kolkata (d) none of these

8. International Plant Genetic Resources Institute (IPGRI) is located in:

(a) Geneva (b) London
(c) Rome (d) India

9. Plant Introduction may lead to entry of new:

(a) Diseases (b) Insects

(c) Weeds (d) All of these

10. Which of the rice variety is an example of primary introduction?

(a) IR 8 (b) IR 20

(c) IR 36 (d) All of these

11. Who coined the term pure line?

(a) Johannsen (b) Louis de Vilmorin

(c) Bridgws (d) none of these

12. Who coined the terms genotype and phenotype?

(a) Johannsen (b) Louis de Vilmorin

(c) Bridgws (d) none of these

13. Who developed the concept of pure line selection?

(a) Johannsen (b) Louis de Vilmorin

(c) Bridgws (d) none of these

14. Pure line variety is:

(a) Homozygous (b) Homogeneous

(c) Highly uniform (d) All of these

15. With which crop the concept of pure line selection was developed?

(a) French bean (b) Corn

(c) Garden pea (d) none of these

16. In a pure line variety variation arises due to:

(a) Out crossing (b) Mechanical admixture

(c) Spontaneous mutations (d) All of these

D. TRUE OR FALSE

1. In cotton, PRS 72 was introduced from USSR.
2. Wheat variety Sonora 64 is an example of primary introduction.
3. Wheat variety Kalyan Sona is an example of secondary introduction.
4. Rice variety IR 8 is an example of primary introduction.
5. Primary introduction is also known as direct introduction.
6. Secondary introduction is also known as indirect introduction.
7. International Plant Genetic Resources Institute (IPGRI) is located in Rome.
8. National Bureau of Plant Genetic Resources is located in New Delhi.

9. Forest Research Institute is located in Dehradun.
10. Botanical Survey of India is located in Kolkata.
11. The concept of pure line theory was developed by Johannsen.
12. Pure line selection is commonly used in self-pollinated species.
13. All the plants of a pure line have exactly the same genotype.
14. Pure line varieties are homozygous and homogeneous populations.
15. Pure lines have narrow genetic base.
16. There is no genetic variation in a pure line
17. In a pure line, selection is ineffective due to lack of genetic variation.
18. The variation within a pure line is entirely due to environmental factors.

E. FILL IN THE BLANKS

1. In cotton, PRS 72 was introduced from ______.
2. Wheat variety Sonora 64 is an example of _____ introduction.
3. Wheat variety Kalyan Sona is an example of ______ introduction.
4. Rice variety IR 8 is an example of ______ introduction.
5. Primary introduction is also known as _____ introduction.
6. Secondary introduction is also known as _____introduction.
7. International Plant Genetic Resources Institute (IPGRI) is located in _____.
8. National Bureau of Plant Genetic Resources is located in _____.
9. Forest Research Institute is located in _______.
10. Botanical Survey of India is located in _______.
11. The concept of pure line theory was developed by ______.
12. Pure line selection is commonly used in _______ species.
13. All the plants of a pure line have exactly the same _____.
14. Pure line varieties are homozygous and _______ populations.
15. Pure lines have _______genetic base.
16. There is no ______ variation in a pure line
17. In a pure line, selection is ineffective due to lack of _____ variation.
18. The variation within a pure line is entirely due to _______ factors.

F. ONE WORD ANSWER

1. A foreign variety which is directly recommended for commercial cultivation.
2. Transposition of crop plants from the place of their cultivation to such areas where they were never grown earlier.

3. Introductions which are immediately adapted to the changed environment.
4. Introductions which take some years for adaptation in changed environment.
5. Introduction that can be used for commercial cultivation without any change in the original genotype.
6. Introduction that can be used as a variety after selection from original genotype or used for transfer of some desirable gene to the cultivated variety.
7. Adaptation of an introduced variety to the new environment.
8. The process of bringing wild and weedy species under human management.
9. Recording observations on highly heritable traits such as colour, shape, hairiness, *etc.*
10. The process that favors survival and further propagation of some plants having more desirable characters than others. It is of two types, *viz.*, natural and artificial.
11. Selection that operates in nature without human interference. It favors those plant characters that are essential for survival (adaptation) of a species.
12. The selection made by human. It favors those characteristics of plants that are related to yield and quality.
13. Selection in favor of intermediate types; also called centripetal selection.
14. Selection in favor of an extreme phenotype such as earliness or lateness, dwarfness or tallness.
15. Selection in favor of two extreme types, *viz.* earliness and lateness or tallness and dwarfness.
16. Selection in one direction for one generation or season and in opposite direction in next generation or season.
17. Progeny of a self-pollinated homozygous plant obtained by selfing.
18. Development of new variety through identification and isolation of a single best plant progeny.

ANSWERS

C. MULTIPLE CHOICE QUESTIONS

Sl.No.	*Answer*	*Sl.No.*	*Answer*
1	(d) All of these	2	(a) NBPGR, New Delhi
3	(b) FRI, Dehradun	4	(c) BSI, Kolkata
5	(a) New Delhi	6	(b) Dehradun
7	(c) Kolkata	8	(c) Rome
9	(d) All of these	10	(d) All of these
11	(a) Johannsen	12	(a) Johannsen
13	(a) Johannsen	14	(d) All of these
15	(a) French bean	16	(d) All of these

D. TRUE OR FALSE

1 to 18 all are true.

E. FILL IN THE BLANKS

Sl.No.	*Answer*	*Sl.No.*	*Answer*
1	USSR	2	Primary
3	Secondary	4	Primary
5	Direct	6	Indirect
7	Rome	8	New Delhi
9	Dehradun	10	Kolkata
11	Johannsen	12	Self-pollinated
13	Genotype	14	Homogeneous
15	Narrow	16	Genetic
17	Genetic	18	Environmental

F. ONE WORD OR SHORT ANSWER

Sl.No.	*Answer*	*Sl.No.*	*Answer*
1	Exotic Variety	2	Plant Introduction
3	Direct Introduction	4	Indirect Introduction
5	Primary introduction	6	Secondary Introduction

Sl.No.	*Answer*	*Sl.No.*	*Answer*
7	Acclimatization	8	Domestication
9	Characterization	10	Selection
11	Natural Selection	12	Artificial Selection
13	Stabilizing Selection	14	Directional Selection
15	Disruptive Selection	16	Cyclic Selection
17	Pure Line	18	Pure line Selection

Chapter 7

Breeding Self-pollinated Crops (Pedigree, Bulk, SSD and Backcross Methods)

A. DEFINITIONS

Backcross: Crossing of F_1 with either of its parents. When F_1 is crossed with homozygous recessive parent, it is called test cross.

Bulk Breeding: A selection procedure which is used in segregating population of self-pollinated species in which material is grown in bulk plot from F_2 to F_5 with or without selection, next generation is grown from bulk seed and individual plant selection is practiced in F_6 or later generations.

Donor Parent: The parent which donates desirable genes; also called non-recurrent parent, because it is used once in crossing program.

Hybridization: Crossing between genetically dissimilar plants. It may involve two genotypes of the same species (inter-varietal hybridization) or two species of the same genus (interspecific hybridization) or two genera of the same family inter-generic hybridization).

Isogenic Lines: Genotypes having single locus difference only.

Mass Pedigree Method: A modified form of pedigree method in which segregating material is handled by bulk (mass) method when conditions are unfavorable for selection and by pedigree method when conditions are favorable for selection.

Multi-lines: The deliberate seed mixtures of isogenic lines, closely related lines or unrelated lines. A variety developed for commercial cultivation from any of these mixtures is known as multiline variety and such procedure is called multiline breeding.

Pedigree Breeding: A selection procedure which is used in segregating population of self-pollinated species and keeps proper record of plants and progeny selected in each generation.

Pedigree: Record of the ancestry of an individual selected plant for its various generations.

Recipient Parent: The parent which receives a desirable character, also called recurrent parent, because it is repeatedly used in backcrossing.

Single Seed Descent: A breeding procedure which is used with segregating population of self-pollinated species in which plants are advanced by selecting single seed per plant from F_2 generation onwards.

B. BACKGROUND INFORMATION

1. There are four hybridization methods which are commonly used for genetic improvement of self-pollinated crops. These are pedigree breeding, bulk breeding, single seed descend and backcross breeding. A brief description of first three methods is presented in this chapter.
2. Hybridization is used when there is limited genetic variability in a species and improvement is not possible by pure line selection and mass selection.
3. Hybridization may involve genotypes of the same species, different species and different genera of the same family.
4. Depending upon the involvement of genotypes, hybridization is of following three types, *viz.* intraspecific, interspecific and inter-generic.
5. Intra-specific hybridization involves two different genotypes of the same species. It is also called inter-varietal hybridization.
6. Interspecific hybridization involves two different species belonging' to the same genus. It is also known as intra-generic hybridization.
7. Inter-generic hybridization involves two species each belonging to different genus of a family.
8. The important objectives of hybridization are: (i) to combine several desirable traits from different sources into a single genotype, (ii) to create vast genetic variability for various economic characters in a population, (iii) to develop superior crop cultivars, and (iv) to develop hybrid varieties for commercial cultivation.
9. The variety developed by pedigree method is homozygous and homogeneous, because it is a progeny of single homozygote.
10. In Rice, varieties Krishna, Ratana, Sabarmati, Padma, Jaya, Bala, Kaveri, *etc.* have been developed by pedigree method.
11. In wheat, varieties K65, K68, NP52, NP120, NP125, NP700, *etc.* have been developed by pedigree method.
12. In cotton, varieties J34, J205, LH372, H655C, SH131, MCU8, MCU9, Sujay, Jayadhar, Raichur 51, Suvin, *etc.* have been developed by pedigree method.

13. In Chick pea, varieties T1, T2, T3, T5, Radhey, *etc.* have been developed by pedigree method.
14. Backcross refers to crossing of *F1* with either of its parents. When the F_1 is crossed with homozygous recessive parent, it is known as test cross.
15. A system of breeding in which repeated backcrosses are made to transfer a specific character to a well-adapted variety for which the variety is deficient is referred to as backcross breeding.
16. Backcross method involves two types of parents, *viz.*, recipient parent and donor parent. The recipient parent is generally a well-adapted high yielding variety of an area which is deficient in one or few characters.
17. Generally 5 to 6 backcrosses are sufficient to retain the genotype of original variety with new character.
18. The basic requirements to start a backcross program are (i) recurrent parent, (ii) donor parent, and (iii) high heritability of the character under transfer.
19. The end product of backcross breeding method is a homozygous arid homogeneous population, because it is the progeny of a single homozygote.
20. In cotton varieties V797, Digvijay, Vijalpa and Kalyan which belong to *Gossypium herbaceum* have been developed by backcross method.
21. The use of multiline cultivars was first suggested in oats by Jensen in 1952. Borlaug and Gibler in 1953 outlined the method for developing multi-lines in wheat.
22. The multiline approach is applicable to self-pollinated species only. Multiline cultivars are commercially used in self/pollinated crops like oat, wheat, soybean, ground nut and many other crops.
23. Multiline cultivars are mixtures of several pure lines. The pure lines may be isogenic lines, closely related lines or unrelated lines.
24. Multiline cultivars have been developed for commercial cultivation in oats, wheat, soybean and peanut in USA. In India three multiline varieties, *viz.* KSML 3, MLKS 11 and KML 7404 have been released in wheat from Punjab. The first two varieties involve 8 closely related lines and the third one involves 9 closely related lines.
25. There are three types of multilines: *viz.* (i) mixtures of isolines, (ii) mixtures of closely related lines, and (iii) mixtures of unrelated or distinctly different genotypes.
26. The development of a multiline consists of four important steps: (i) selection of recurrent parent, (ii) selection of donor parents, (iii) transfer of resistant genes into recurrent parent, and (iv) mixing seed of the isogenic lines.

27. The resistant genes are transferred from donor parents to the recurrent parent through a series of several separate backcross program. Generally 4-5 backcrosses are sufficient to retain the genotype of recurrent parent with added resistance in the backcross derivatives.

C. MULTIPLE CHOICE QUESTIONS

1. **Breeding methods involving hybridization refer to:**
   (a) Pedigree method (b) Bulk method
   (c) Back cross method (d) All of these
2. Which of the following breeding methods does not involve hybridization?
   (a) Pedigree method (b) Mutation breeding
   (c) Back cross method (d) All of these
3. **Which of the following is called evolutionary method of breeding?**
   (a) Pedigree method (b) Mutation breeding
   (c) Bulk method (d) All of these
4. **Which of the following methods throws transgressive segregants?**
   (a) Pedigree method (b) Single seed descent
   (c) Bulk method (d) All of these
5. **Which of the following methods involves progeny testing?**
   (a) Pedigree method (b) Single seed descent
   (c) Bulk method (d) All of these
6. **A pedigree bred variety has:**
   (a) High uniformity (b) Narrow genetic base
   (c) Narrow adaptation (d) All of these
7. **A pedigree bred variety is:**
   (a) High uniform (b) Homozygous
   (c) Homogeneous (d) All of these
8. **Who developed the concept of bulk breeding method?**
   (a) Nilsson Ehle (b) Goulden
   (c) Harland (d) All of these
9. **Who developed the concept of single seed descent method (SSD)?**
   (a) Nilsson Ehle (b) Goulden
   (c) Harland (d) All of these

10. Who developed Mass Pedigree method?

(a) Nilsson Ehle (b) Goulden

(c) Harrington (d) All of these

11. Who coined the term test cross?

(a) Morgan (b) Bidges

(c) Darlington (d) None of these

12. Which breeding method is used for transfer of oligogenic characters?

(a) Bulk breeding (b) Back cross

(c) Pedigree method (d) None of these

13. In which breeding method repeated crossing is involved?

(a) Bulk breeding (b) Back cross

(c) Pedigree method (d) None of these

14. In which breeding method recurrent and non-recurrent parents are used?

(a) Bulk breeding (b) Back cross

(c) Pedigree method (d) None of these

15. Which breeding method is used for developing multi-lines?

(a) Bulk breeding (b) Back cross

(c) Pedigree method (d) None of these

16. Which breeding method is used transfer of disease resistance?

(a) Bulk breeding (b) Back cross

(c) Pedigree method (d) None of these

17. Which breeding method is used transfer of male sterility?

(a) Bulk breeding (b) Back cross

(c) Pedigree method (d) None of these

18. Multi-line is a mixture of several:

(a) Pure lines (b) Inbred lines

(c) Isogenic lines (d) None of these

19. KSML 3 is a multi-line variety of which crop?

(a) Wheat (b) Rice

(c) Barley (d) None of these

20. From which research center the KSML 3 variety of wheat was developed?

(a) HAU, Hisar (b) PAU, Ludhiana

(c) IWBRI, Karnal (d) None of these

D. TRUE OR FALSE

1. The ancestry of an individual selected plant is called pedigree.
2. Pedigree method is widely used for the improvement of self-pollinated species.
3. It is generally used when both the parents that are used in the hybridization are well adapted.
4. In pedigree method, only human selection or artificial selection is used.
5. Pedigree method takes 14-15 years for development of new crop cultivar.
6. Mass pedigree method is a modification of pedigree method. It was proposed by Harrington in 1937.
7. Bulk breeding method is also referred to as evolutionary method of crop improvement.
8. Bulk breeding method was developed by Nilsson Ehle in 1908.
9. SSD method was suggested by Goulden (1939) for advancing segregating generation of self-pollinated crops.
10. In India, only one variety "Narendra Rai" has been developed in Brown Mustard by bulk method.
11. SSD method was applied by Grafius (1965) in oats, Brim (1966) in soybean.
12. Backcross method is more commonly used for transfer of oligogenic traits.
13. It is more successful when the character has high heritability.
14. Oligogenic characters have high heritability than polygenic traits.
15. The parent which receives a desirable character is known as recipient parent.
16. The recipient parent is repeatedly used in the backcross method.
17. Recipient parent is also called as recurrent parent.
18. The parent which donates the desirable character is known as donor parent.
19. Donor parent is also known as non-recurrent parent.
20. Backcross method is used to rectify the specific defect of an adapted variety.
21. Isogenic lines have single gene difference.
22. KSML 3 is a multiline variety of Wheat.
23. Multiline variety KSML 3 was developed from PAU, Ludhiana.

E. FILL IN THE BLANKS

1. The ancestry of an individual selected plant is called ______.
2. Pedigree method is widely used for the improvement of ______ species.

3. It is generally used when both the parents that are used in the hybridization are well ______.
4. In pedigree method, only ______ selection is used.
5. Pedigree method takes ______ years for development of new crop cultivar.
6. Mass pedigree method is a modification of pedigree method. It was proposed by ________.
7. Bulk breeding method is also referred to as ______method of crop improvement.
8. Bulk breeding method was developed by ______.
9. SSD method was suggested by ______for advancing segregating generation of self-pollinated crops.
10. In India, only one variety "Narendra Rai" has been developed in Brown Mustard by ______ method.
11. SSD method was applied by ______ in oats.
12. Backcross method is more commonly used for transfer of ______ traits.
13. It is more successful when the character has ______ heritability.
14. Oligogenic characters have high heritability than ______traits.
15. The parent which receives a desirable character is known as ______
16. The ______parent is repeatedly used in the backcross method.
17. Recipient parent is also called as ______ parent.
18. The parent which donates the desirable character is known as ______.
19. Donor parent is also known as non-recurrent parent.
20. Backcross method is used to rectify the specific defect of an ______.
21. Isogenic lines have ______ gene difference.
22. KSML 3 is a multiline variety of ______.
23. Multiline variety KSML 3 was developed from ______.

F. ONE WORD OR SHORT ANSWER

1. Crossing between genetically dissimilar plants.
2. Record of the ancestry of an individual selected plant for its various generations.
3. A selection procedure which is used in segregating population of self-pollinated species and keeps proper record of plants and progeny selected in each generation.
4. A modified form of pedigree method in which segregating material is handled by bulk (mass) method when conditions are unfavorable for selection and by pedigree method when conditions are favorable for selection.

5. A selection procedure which is used in segregating population of self-pollinated species in which material is grown in bulk plot from F_2 to F_5 with or without selection, next generation is grown from bulk seed and individual plant selection is practiced in F_6 or later generations.
6. A breeding procedure which is used with segregating population of self-pollinated species in which plants are advanced by selecting single seed per plant from F_2 generation onwards.
7. Crossing of F_1 with either of its parents. When F_1 is crossed with homozygous recessive parent, it is called test cross.
8. The parent which donates desirable genes; also called non-recurrent parent, because it is used once in crossing program.
9. The parent which receives a desirable character, also called recurrent parent, because it is repeatedly used in backcrossing.
10. The deliberate seed mixtures of isogenic lines, closely related lines or unrelated lines. A variety developed for commercial cultivation from any of these mixtures is known as multiline variety and such procedure is called multiline breeding.
11. Genotypes having single locus difference only.

ANSWERS

C. MULTIPLE CHOICE QUESTIONS

Sl.No.	*Answer*	*Sl.No.*	*Answer*
1	(d) All of these	2	(b) Mutation breeding
3	(c) Bulk method	4	(a) Pedigree method
5	(a) Pedigree method	6	(d) All of these
7	(d) All of these	8	(a) Nilsson Ehle
9	(b) Goulden	10	(c) Harrington
11	(b) Bidges	12	(b) Back cross
13	(b) Back cross	14	(b) Back cross
15	(b) Back cross	16	(b) Back cross
17	(b) Back cross	18	(c) Isogenic lines
19	(a) Wheat	20	(b) PAU, Ludhiana

D. TRUE OR FALSE

1 to 23 all are true.

E. FILL IN THE BLANKS

1	Pedigree	2	Self-pollinated
3	Well adapted	4	Human
5	14-15	6	Harrington
7	Evolutionary	8	Nilsson Ehle
9	Goulden	10	Bulk
11	Grafius	12	Oligogenic
13	High	14	Polygenic
15	Recipient parent	16	Recipient
17	Recurrent	18	Donor parent
19	Non-recurrent	20	Adapted variety
21	Single	22	Wheat
23	PAU, Ludhiana		

F. ONE WORD OR SHORT ANSWER

Sl.No.	*Answer*	*Sl.No.*	*Answer*
1	Hybridization	2	Pedigree
3	Pedigree Breeding	4	Mass Pedigree Method
5	Bulk Breeding	6	Single Seed Descent
7	Backcross	8	Donor Parent
9	Recipient Parent	10	Multi-lines
11	Isogenic Lines		

Chapter 8

Breeding Cross-pollinated Crops (Mass, Progeny and Ear to Row Selections)

A. DEFINITIONS

Line Breeding: A system of breeding in which a number of genotypes with superior performance for several characters are composed to form a variety.

Mass Selection: A method of crop improvement in which individual desirable plants are selected on the basis of phenotype from a mixed population, their seeds are bulked and used to grow the next generation **(positive mass selection).** Sometimes, only undesirable off type plants are removed from the field and rest are allowed to grow further **(negative mass selection).**

Progeny Selection: A selection procedure in which superior plants are selected from a heterogeneous population on the basis of the performance of their progeny.

B. BACKGROUND INFORMATION

1. In mass selection, individual plants are selected on the basis of phenotype from a mixed population their seeds are bulked and used to grow the next generation.
2. There are two types of mass selection, *viz.* (1) positive mass selection, and (2) negative mass selection.
3. The genetic worth of an individual is assessed by progeny test: The test of the genotypic value of an individual based on the performance of its progeny is called progeny test.
4. The progeny test was developed by Louis de Vilmorin, hence it is also known as Vilmorin principle.

5. The progeny test is useful in two ways, *viz.* (i) In understanding whether a plant is true breeding (homozygous) or segregating (heterozygous) for a particular character, and (ii) in the assessment of the breeding value of a plant. If the selected plant is really superior, its progeny will also exhibit superior performance and vice versa.
6. Selection of plants from a diverse population on the basis of their progeny test is called progeny selection.
7. In cross pollinated species, three types of materials, *viz.* (i) open pollinated seeds, (ii) self-seeds, and (iii) top cross or test cross seeds of selected plants can be used in producing the progenies for testing.
8. In cross pollinated species various breeding materials such as an open pollinated variety, composite, synthetic variety, a mass selected variety, a single cross hybrid, a three-way cross hybrid or double cross hybrid can be used as base material to start ear to row selection.
9. There is genetic variation in the variety developed by ear to row selection, because such variety is a mixture of several homozygotes and heterozygotes. Thus, selection is effective in such variety.
10. In cross pollinated species, a variety developed by line breeding approach consists of several homo and heterozygotes and thus has genetic diversity.

C. MULTIPLE CHOICE QUESTIONS

1. **The concept of ear to row selection was developed by:**

(a) Vilmorin (b) Hopkins

(c) Bridges (d) All of these

2. **The concept of progeny test was developed by:**

(a) Vilmorin (b) Hopkins

(c) Bridges (d) All of these

3. **In cross-pollinated crops, which material can be used to start Ear to Row selection?**

(a) An OP variety (b) Composite Variety

(c) Synthetic variety (d) All of these

4. **In cross-pollinated crops, which material can be used to start Ear to Row selection?**

(a) Mass selected variety (b) Composite Variety

(c) Synthetic variety (d) All of these

5. **In cross-pollinated crops, which material can be used to start Ear to Row selection?**

(a) Single cross hybrid (b) Three-way cross hybrid

(c) Double cross hybrid (d) All of these

6. **Mass selection is based on:**
   (a) Phenotype (b) Progeny test
   (c) Breeding value (d) All of these

7. **Progeny selection is based on:**
   (a) Phenotype (b) Progeny test
   (c) Breeding value (d) All of these

8. **Who suggested the grid method mass selection?**
   (a) East (b) Gardner
   (c) Allard (d) None of these

9. **A variety developed by Ear to Row selection consists of several:**
   (a) Homozygotes (b) Heterozygotes
   (c) Homo and heterozygotes (d) None of these

10. **A variety developed by line breeding consists of several:**
    (a) Homozygotes (b) Heterozygotes
    (c) Homo and heterozygotes (d) None of these

D. TRUE OR FALSE

1. In mass selection, individual plants are selected on the basis of phenotype.
2. Mass selection is applicable to both self and cross pollinated species.
3. In self-pollinated crops, a mass selected variety is homozygous but heterogeneous.
4. There is heritable variation in the mass selected varieties.
5. Selection is effective in case of mass selected varieties.
6. In mass selection, there is no control on the pollination.
7. The progeny test was developed by Louis de Vilmorin.
8. The progeny test is also known as Vilmorin principle.
9. The concept of ear to row selection was developed by Hopkins in 1908 working with maize crop in USA.
10. The ear to row selection is a form of progeny selection.
11. The ear to row selection was first used by Hopkins in maize.

E. FILL IN THE BLANKS

1. In mass selection, individual plants are selected on the basis of ______.
2. Mass selection is ______to both self and cross pollinated species.
3. In self-pollinated crops, a mass selected variety is homozygous but ______.

4. There is ______ variation in the mass selected varieties.
5. Selection is ______ in case of mass selected varieties.
6. In mass selection, there is no control on the ______.
7. The progeny test was developed by ______.
8. The progeny test is also known as ______.
9. The concept of ear to row selection was developed by ______.
10. The ear to row selection is a form of ______ selection.
11. The ear to row selection was first used by ______ in maize.

F. ONE WORD OR SHORT ANSWER

1. A method of crop improvement in which individual desirable plants are selected on the basis of phenotype from a mixed population.
2. Selection of desirable plants from a mixed population.
3. Removal of undesirable off type plants from a mixed population.
4. A selection procedure in which superior plants are selected from a heterogeneous population on the basis of the performance of their progeny.
5. A system of breeding in which a number of genotypes with superior performance for several characters are composed to form a variety.

ANSWERS

C. MULTIPLE CHOICE QUESTIONS

Sl.No.	*Answer*	*Sl.No.*	*Answer*
1	(b) Hopkins	2	(a) Vilmorin
3	(d) All of these	4	(d) All of these
5	(d) All of these	6	(a) Phenotype
7	(b) Progeny test	8	(b) Gardner
9	(c) Homo and heterozygotes	10	(c) Homo and heterozygotes

D. TRUE OR FALSE

1 to 11 all are true.

E. FILL IN THE BLANKS

Sl.No.	*Answer*	*Sl.No.*	*Answer*
1	Phenotype	2	Applicable
3	Heterogeneous	4	Heritable
5	Effective	6	Pollination
7	Louis de Vilmorin	8	Vilmorin principle
9	Hopkins	10	Progeny
11	Hopkins		

F. ONE WORD OR SHORT ANSWER

Sl.No.	*Answer*	*Sl.No.*	*Answer*
1	Mass Selection	2	Positive mass selection
3	Negative mass selection	4	Progeny Selection
5	Line Breeding		

Chapter 9

Breeding Vegetatively Propagated Crops [Clonal Selection]

A. DEFINITIONS

Clonal Selection: A procedure of selecting superior clones from the mixed population of asexually propagated crops such as sugarcane, potato, *etc.*

Clone: Progeny of a single plant obtained by asexual reproduction.

B. BACKGROUND INFORMATION

1. Those crops which are propagated asexually or by vegetative means are known as asexually propagated or vegetatively propagated or clonal crops.
2. Some examples of asexually propagated crop plants are horticultural Crops such as Banana, mango, citrus, apple, pears, peaches, loquat, litchi, *etc.*
3. Asexually propagated agricultural crops include Sugarcane, potato, sweet potato, *etc.*
4. The main reasons of asexual reproduction are: (i) reduced flowering and seed set, (ii) non flowering in many cases, (iiii) to avoid inbreeding depression in certain crops, and (iv) apomixis in some species.
5. There are three main sources of variation in a clone, *viz.*, bud mutations, mechanical mixtures, and occasional sexual reproduction.
6. Clones have several important features such as: (i) homogeneous constitution, (ii) heterozygosity, (iii) vigorous growth, (iv) wider adaptation, *etc.*
7. In a clonal variety, the variation is entirely due to environmental factors, because all the genotypes of such varieties are identical.

8. Clonal selection is relevant to vegetatively propagated or asexually propagated crops such as sugarcane, potato, sweet potato and some fruit crops.
9. In India, clonal selection has been successfully used for developing new varieties in potato, sugarcane, banana, citrus and grapes.
10. Varieties Kufri Red and Kufri safed in potato; Ko 11, Ko 22 and Neelam in mango; and Bombay green, Pride monthan and High gate in banana have been developed by clonal selection.
11. Several varieties of sugarcane (Co 541, CoS 510, Co 1148, Co 1158 *etc.*) and potato (Kufri, Sinduri, Kufri Kuber, Kufri Kundan, Kufri Chamatkar, *etc.*) have been developed by interspecific hybridization followed by clonal selection.

C. MULTIPLE CHOICE QUESTIONS

1. Which of the following is asexually propagated?

(a) Banana (b) Mango

(c) Citrus (d) All of these

2. Which of the following is asexually propagated?

(a) Apple (b) Pears

(c) Peaches (d) All of these

3. Which of the following is asexually propagated?

(a) Wheat (b) Potato

(c) Sorghum (d) All of these

4. Which of the following is asexually propagated?

(a) Sugarcane (b) Barley

(c) Sorghum (d) All of these

5. Kufri Red is a variety of which crop?

(a) Wheat (b) Potato

(c) Sorghum (d) All of these

6. Kufri Safed is a variety of which crop?

(a) Wheat (b) Potato

(c) Sorghum (d) All of these

7. Neelam is a variety of which crop?

(a) Mango (b) Potato

(c) Sorghum (d) All of these

8. **Bombay green is a variety of which crop?**
   (a) Mango (b) Potato
   (c) Banana (d) All of these

9. **Kufri Kundan is a variety of which crop?**
   (a) Mango (b) Potato
   (c) Banana (d) All of these

10. **Where is Central Potato Research Institute located?**
   (a) Shimla (b) Nainital
   (c) Gwalior (d) All of these

D. TRUE OR FALSE

1. Potato propagates by tubers.
2. Sugarcane propagates by stem cutting.
3. Sweet Potato propagates by root cuttings.
4. The progeny of a clone is genetically identical.
5. Clones are related to asexually propagated species.
6. Clones are heterozygous and homogeneous.
7. Clones are maintained by asexual reproduction.
8. Clones are stable and are not prone to segregation.
9. Exact reconstitution of a clone is not possible.
10. Clonal selection is useful in conserving heterosis.

E. FILL IN THE BLANKS

1. Potato propagates by ______.
2. Sugarcane propagates by ______.
3. Sweet Potato propagates by ______.
4. The progeny of a clone is genetically ______.
5. Clones are related to asexually propagated ______.
6. Clones are heterozygous and ______.
7. Clones are maintained by ______.
8. Clones are stable and are not prone to ______.
9. Exact reconstitution of a clone is ______.
10. Clonal selection is useful in conserving ______.

F. ONE WORD OR SHORT ANSWER

1. Progeny of a single plant obtained by asexual reproduction.
2. A procedure of selecting superior clones from the mixed population of asexually propagated crops such as sugarcane, potato, *etc.*

ANSWERS

C. MULTIPLE CHOICE QUESTIONS

Sl.No.	*Answer*	*Sl.No.*	*Answer*
1	(d) All of these	2	(d) All of these
3	(b) Potato	4	(a) Sugarcane
5	(b) Potato	6	(b) Potato
7	(a) Mango	8	(c) Banana
9	(b) Potato	10	(a) Shimla

D. TRUE OR FALSE

1 to 10 all are true.

E. FILL IN THE BLANKS

Sl.No.	*Answer*	*Sl.No.*	*Answer*
1	Tubers	2	Stem cuttings
3	Root cuttings	4	Identical
5	Species/ crops	6	Homogeneous
7	Asexual reproduction	8	Segregation
9	Not possible	10	Heterosis

F. ONE WORD OR SHORT ANSWER

Sl.No.	*Answer*	*Sl.No.*	*Answer*
1	Clone	2	Clonal Selection

Chapter 10

Transgenic Breeding

A. DEFINITIONS

***Agrobacterium tumifaciens*:** A soil borne bacterium which is widely used for development of transgenic plants in different crop plants. It is a biological method of foreign gene transfer into host cells.

Callus: A mass of unorganized regenerated cells in culture medium.

Explant: The plant part which is used for regeneration into whole plant in tissue culture.

Gene Cloning: The process of making several identical copies of a gene in genetic engineering.

Genetic Engineering: Isolation, introduction and expression of foreign DNA in Plants and animals.

Genetic Transformation: The process of transfer, integration and expression of transgene (foreign gene) in an organism.

Micro-Injection: A mechanical method of gene transfer in which foreign DNA is delivered into the host cells by microscopic needles also known as micro-injection.

Particle Bombardment: A method of foreign gene transfer which involves high velocity metal particles (usually gold or tungsten) or metal particles mediated method of genetic transformation. It is also known as biolistic micro projectile bombardment, micro ballistics and particle acceleration.

Plant Biotechnology: Combined study of plant tissue culture and genetic engineering or recombinant DNA technology.

Plant Tissue Culture: Growth of tissues of living plants in a suitable culture medium (*in vitro*). It includes cell culture, protoplast culture, organ culture and meristem culture.

Plasmid: An extra-chromosomal genetic element which is found in the bacteral cell, can replicate independently of chromosomal DNA and is not essential for normal growth and development of bacterium.

Protoplasts: Naked cells without cell wall.

Somaclonal variation: The variability generated by the use of tissue culture technique.

Transformants: Cells or tissues of an organism which have undergone genetic transformation; resultants of genetic transformation.

Transgenes: Foreign genes or modified genes of the same species which are used for the development of transgenic individuals. Such gene may be from the same species (in modified form), related wild species, unrelated species and microbes (bacteria, fungi and viruses).

Transgenic Breeding: Genetic improvement of crop plants, domestic animals and useful micro-organisms through biotechnology.

Transgenic: Genetically engineered organisms or organisms developed by the techniques of genetic engineering. It may be a plant, an animal or micro-organism.

B. BACKGROUND INFORMATION

1. The technique or process of developing transgenic individuals is referred to as transgenic technology.
2. It is also known as gene technology, transformation technology, genetic engineering and recombinant DNA technology.
3. The crop plants which are developed through the application of plant biotechnology or plant genetic engineering are called transgenic plants or genetically modified plants or genetically engineered plants.
4. In broad sense transgenic plants can be called as genetically modified plants. But in strict sense, they are genetically engineered plants. However, the term GM crops or GM plants or GM organisms has been very popular and is widely used the world over.
5. The GM plants have been developed to solve specific problems which could not be solved easily through conventional plant breeding methods.
6. The genetically engineered plants are called transgenic plants. Transgenic plants contain transgene or foreign gene of special significance.
7. Transgenic plants are developed through plant biotechnology.
8. In the development of transgenic plants, the sexual process is not involved.
9. The transgenic plants are recovered at a very low frequency.
10. Transgenic plants are developed to solve those problems which cannot be solved by conventional plant breeding techniques.

11. Transgenic technology has several advantages such as rapid method, free gene transfer, single gene transfer, direct gene transfer and solution to difficult problems.
12. Important applications of transgenic technology include improvement in (i) biotic resistance, (ii) abiotic resistance, (iii) herbicide resistance, (iv) quality of food products and (v) development of novel traits such as golden rice and male sterility.
13. Plant biotechnology has played significant role in (i) improving resistance [to insects, diseases, drought, salinity, frost, cold, *etc.* (ii) food quality and (iii) inducing male sterility in some crops.
14. Insect resistance includes bollworm resistance in cotton, European borer resistance in maize and Colorado beetle resistance in Potato.
15. Resistance to these insects has been transferred from *Bacillus thuringiensis.*
16. Salinity resistant tomato has been developed at the University of California and at the University of Toronto.
17. Herbicide resistance has been achieved in cotton, Flax, Soybean, Wheat, Potato, Tomato, Rapeseed, *etc.*
18. The gene for herbicide resistance has been transferred from Steptomyces to wheat, potato and tomato; and from microbial gene in cotton, soybean and linseed.
19. In potato starch contents have been increased by transferring gene from human intestine bacteria [*Escherichia coli*].
20. In potato, the high protein gene has been transferred from Amaranth and serum albumin gene from human.
21. In alfalfa ova-albumin gene has been transferred from chicken.
22. In rice, high carotene content gene has been transferred from Daffodils and Golden rice has been developed.
23. The golden rice is rich in vitamin A content. Two genes have been transferred from Daffodils and one from bacterial species.
24. In tomato, anti-cracking gene has been transferred from winter flounder fish.
25. In tobacco, freezing resistant gene has been transferred from winter flounder fish and cold resistant gene from *Arabidopsis thaliana.*
26. In rapeseed, male sterility has been developed through the application of agricultural biotechnology. The *barnase* and *barstar* system has been developed in this crop.
27. The male sterility gene has been transferred from *Bacillus amyloliquefaciens.*
28. Probable risks of plant biotechnology include such as adverse effects on human and animal health, development of resistance in some weeds against herbicides, ethical issues and development of resistance in insects and diseases against transgenic plants.

C. MULTIPLE CHOICE QUESTIONS

1. **Who coined the term protoplast?**
   (a) Ingo Potrikus (b) Gordon and Ruddle
   (c) Hanstein (d) none of these
2. **Who coined the term transgenic?**
   (a) Ingo Potrikus (b) Gordon and Ruddle
   (c) Hanstein (d) none of these
3. **Who first developed Golden rice?**
   (a) Ingo Potrikus (b) Gordon and Ruddle
   (c) Hanstein (d) none of these
4. **Who first discovered barnase and barsrar system in rapeseed?**
   (a) Mariani *et al.* (b) Gordon and Ruddle
   (c) Hanstein (d) none of these
5. **Who first developed Golden rice 2?**
   (a) Ingo Potrikus (b) Gordon and Ruddle
   (c) Syngenta Company (d) none of these
6. **In which year Golden rice was developed?**
   (a) 1980 (b) 1981
   (c) 2000 (d) 2005
7. **In which year Golden rice 2 was developed?**
   (a) 1980 (b) 1981
   (c) 2000 (d) 2005
8. **What is the source of barnase and barstar in rapeseed?**
   (a) Daffodils (b) Amranth
   (c) *Escherichia coli* (d) *Bacillus amyloliquefaciens.*
9. **From which source anti-cracking gene has been transferred in tomato?**
   (a) Daffodils (b) Amranth
   (c) *Escherichia coli* (d) Winter flounder fish
10. **From which source freezing resistant gene has been transferred in tobacco?**
    (a) Daffodils (b) Amranth
    (c) *Escherichia coli* (d) Winter flounder fish

11. From which source high protein gene has been transferred in potato?

(a) Daffodils
(b) Amranth
(c) *Escherichia coli*
(d) Winter flounder fish

12. From which source high starch gene has been transferred in potato?

(a) Daffodils
(b) Amranth
(c) *Escherichia coli*
(d) Winter flounder fish

D. MARK TRUE OR FALSE

1. Hanstein first used the term protoplast in 1980.
2. Gordon and Ruddle first used the term transgenic in 1981.
3. Ingo Potrykus and Peter Beyer first developed the Golden Rice in 2000.
4. Syngenta Company developed Golden Rice 2 in 2005.
5. Golden Rice 2 has 23 times more beta-carotene than the original variety of golden rice.
6. Mariani *et al.*, first reported the barnase and barstar system of male sterility in rapeseed in 1990.
7. In alfalfa ova-albumin gene has been transferred from chicken.
8. In Golden rice, high carotene content gene has been transferred from Daffodils.
9. The *barnase* and *barstar* system has been developed in rapeseed.
10. In Rapeseed, the male sterility gene has been transferred from *Bacillus amyloliquefaciens.*
11. In tomato, anti-cracking gene has been transferred from winter flounder fish.
12. In tobacco, freezing resistant gene has been transferred from winter flounder fish.
13. In tobacco, cold resistant gene has been transferred from *Arabidopsis thaliana.*
14. In potato high starch contents gene have been transferred from Escherichia coli.
15. In potato, the high protein gene has been transferred from Amaranth.

E. FILL IN THE BLANKS

1. __________ first used the term protoplast in 1980.
2. ___________ first used the term transgenic in 1981.
3. Ingo Potrykus and Peter Beyer first developed the Golden Rice in __________.

4. Syngenta Company developed Golden Rice 2 in ________.
5. Golden Rice 2 has __________ more beta-carotene than the original variety of golden rice.
6. __________ first reported the barnase and barstar system of male sterility in rapeseed in 1990.
7. In alfalfa ova-albumin gene has been transferred from __________.
8. In Golden rice, high carotene content gene has been transferred from ________.
9. The *barnase* and *barstar* system has been developed in ________.
10. In Rapeseed, male sterility gene has been transferred from __________
11. In tomato, anti-cracking gene has been transferred from __________.
12. In tobacco, freezing resistant gene has been transferred from __________.
13. In tobacco, cold resistant gene has been transferred from __________.
14. In potato high starch contents gene have been transferred from __________.
15. In potato, the high protein gene has been transferred from __________.

F. ONE WORD OR SHORT ANSWER

1. Foreign genes or modified genes of the same species which are used for the development of transgenic individuals. Such gene may be from the same species (in modified form), related wild species, unrelated species and microbes (bacteria, fungi and viruses).
2. Genetically engineered organisms or organisms developed by the techniques of genetic engineering. It may be a plant, an animal or micro-organism.
3. The process of transfer, integration and expression of transgene (foreign gene) in an organism.
4. Cells or tissues of an organism which have udergone genetic transformation; resultants of genetic transformation.
5. An extra-chromosomal genetic element which is found in the bacteral cell, can replicate independently of chromosomal DNA and is not essential for normal growth and development of bacterium.
6. Combined study of plant tissue culture and genetic engineering or recombinant DNA technology.
7. Growth of tissues of living plants in a suitable culture medium (*invitro*). It includes cell culture, protoplast culture, organ culture and meristem culture.
8. Isolation, introduction and expression of foreign DNA in Plants and animals.

9. Naked cells without cell wall.

   Callus A mass of unorganized regenerated cells in culture medium.
10. The variability generated by the use of tissue culture technique.
11. The plant part which is used for regeneration into whole plant in tissue culture.
12. The process of making several identical copies of a gene in genetic engineering.
13. A method of foreign gene transfer which involves high velocity metal particles (usually gold or tungsten) or metal particles mediated method of genetic transformation. It is also known as biolistic micro projectile bombardment, micro ballistics and particle acceleration.
14. A soil borne bacterium which is widely used for development of transgenic plants in different crop plants. It is a biological method of foreign gene transfer into host cells.
15. A mechanical method of gene transfer in which foreign DNA is delivered into the host cells by microscopic needles also known as micro-injection.
16. Genetic improvement of crop plants, domestic animals and useful micro-organisms through biotechnology.

ANSWERS

C. MULTIPLE CHOICE QUESTIONS

Sl.No.	*Answer*	*Sl.No.*	*Answer*
1	(c) Hanstein	2	(b) Gordon and Ruddle
3	(a) Ingo Potrikus	4	(a) Mariani *et al.*
5	(c) Syngenta Company	6	(c) 2000
7	(d) 2005	8	(d) *Bacillus amyloliquefaciens.*
9	(d). Winter flounder fish	10	(d). Winter flounder fish
11	(b) Amranth	12	(c) *Escherichia coli*

D. TRUE OR FALSE

1 to 15 all are true.

E. FILL IN THE BLANKS

Sl.No.	*Answer*	*Sl.No.*	*Answer*
1	Hanstein	2	Gordon and Ruddle
3	2000	4	2005
5	23 times	6	Mariani *et al.*
7	Chicken	8	Daffodils
9	Rapeseed	10	*Bacillus amyloliquefaciens.*
11	Winter flounder fish	12	Winter flounder fish
13	*Arabidopsis thaliana*	14	*Escherichia coli*
15	Amaranth		

F. ONE WORD OR SHORT ANSWER

Sl.No.	*Answer*	*Sl.No.*	*Answer*
1	Transgenes	2	Transgenic
3	Genetic Transformation	4	Transformants
5	Plasmid	6	Plant Biotechnology
7	Plant Tissue Culture	8	Genetic Engineering
9	Protoplasts	10	Somaclonal variation
11	Explant	12	Gene Cloning
13	Particle Bombardment	14	*Agrobacterium tumifaciens*
15	Micro-Injection	16	Transgenic Breeding

Chapter 11

SMART Breeding [Marker Assisted Selection]

A. DEFINITIONS

Biochemical Marker: Markers that are related to variation in proteins and amino acid banding pattern.

Cytological Marker: Markers that are related to variation in chromosome number, shape and size.

DNA Marker: A gene or other fragment of DNA whose location in the genome is known. A genetic marker is a known DNA sequence.

Epistasis: The interaction between genes. Epistasis takes place when the action of one gene is modified by one or several other genes.

Genotyping: The process of determining the genotype of an individual with a biological assay.

Locus (plural loci): A fixed position on a chromosome, such as the position of a gene or a biomarker (genetic marker).

Marker assisted selection: Selection for specific alleles (which affect a trait of interest) using genetic markers.

Marker: Any genetic element (locus, allele, DNA sequence or chromosome feature) which can be readily detected by phenotype, cytological or molecular techniques, and used to follow a chromosome or chromosomal segment during genetic analysis.

Molecular Breeding: Improvement of crop plants for various economic characters through indirect selection for linked molecular markers.

Morphological Marker: In plant breeding, markers that are related to variation in shape, size, color and surface of various plant parts.

Phenotype: Any observable characteristic of an organism, such as its morphology, development, biochemical or physiological properties, or behavior.

Pleiotropy: Influence of a single gene on multiple phenotypic traits.

Polymorphism: Presence of multiple alleles of a gene within a population, usually expressing different phenotypes

B. BACKGROUND INFORMATION

1. Marker characters refer to those traits that can be easily identified.
2. Marker assisted selection is also known as marker aided selection (MAS). It is a process whereby a marker (morphological, biochemical or one based on DNA/RNA) is used for indirect selection of a specific trait.
3. Marker assisted selection is based on genetic markers.
4. Marker assisted selection (MAS) is indirect selection process where a trait of interest is selected on a marker linked to it. MAS can be used as an aid to traditional phenotypic-pedigree-based selection systems.
5. MAS can be useful for traits that are difficult to measure, exhibit low heritability, and/or are expressed late in development.
6. Situations in which marker assisted selection is useful include (i) late expression of the trait, (ii) trait controlled by recessive gene, (iii) resistance to biotic stresses, and (iv) presence of epistasis.
7. The marker assisted selection consists of two important steps, *viz.* (i) mapping of gene, and (ii) use of this information for marker assisted selection.
8. Genetic markers have been used in diversity analysis, parentage detection, DNA fingerprinting, and prediction of hybrid performance.
9. Molecular markers are useful in indirect selection processes, enabling manual selection of individuals for further propagation.
10. MAS have also used in backcrossing and gene pyramiding.
11. MAS have several advantages such as (i) high speed, (ii) consistency of results, (iii) bio-safety, (iv) high efficiency, and (v) QTL mapping.
12. The results of MAS are consistent, because use of markers eliminates the impact of environmental variation that often complicates phenotypic evaluation.
13. MAS have very high efficiency. Screening of progeny in early stages allows a breeder to reject undesirable genotypes from the program more quickly.
14. DNA markers permit mapping of polygenic traits or quantitative trait loci [QTL] which is not possible through conventional plant breeding techniques.
15. Main limitations of MAS are (i) expensive technique, (ii) requires technical skill, (iii) a laborious work, and (iv) may sometimes lead to health hazards.

16. Some of the DNA marker techniques involve use of radioactive isotopes in labeling of DNA, which may lead to serious health hazards.

C. MULTIPLE CHOICE QUESTIONS

1. **Genetic markers can be used in:**
   (a) Diversity analysis (b) Parentage detection
   (c) DNA fingerprinting (d) All of these

2. **Markers include:**
   (a) Morphological (b) Molecular
   (c) Biochemical (d) All of these

3. **Marker Assisted Selection is based on which marker?**
   (a) Morphological (b) Molecular
   (c) Biochemical (d) All of these

4. **Advantages of MAS include:**
   (a) High speed (b) Consistent results
   (c) High efficiency (d) All of these

5. **MAS can be useful for traits that are:**
   (a) Difficult to measure (b) Exhibit low heritability
   (c) Expressed late (d) All of these

6. **MAS can be useful for traits that are governed by or exhibit:**
   (a) Recessive genes (b) Low heritability
   (c) Presence of Epistasis (d) All of these

7. **Main limitations of MAS are:**
   (a) Expensive technique (b) High technical skill
   (c) Laborious work (d) All of these

8. **Which type of markers are related to variation in shape, size, color and surface of various plant parts?**
   (a) Morphological (b) Molecular
   (c) Biochemical (d) All of these

9. **Which type of markers are related to variation in proteins and amino acid banding pattern?**
   (a) Morphological (b) Molecular
   (c) Biochemical (d) All of these

10. Which types of markers are related to variation in chromosome number, shape and size?

(a) Morphological (b) Molecular

(c) Biochemical (d) Cytological

D. MARK TRUE OR FALSE

1. Marker characters refer to those traits that can be easily identified.
2. Marker assisted selection is also known as marker aided selection (MAS).
3. Marker assisted selection is based on genetic markers.
4. MAS can be used as an aid to traditional phenotypic selection.
5. The results of MAS are consistent due to little environmental variation
6. MAS can be useful for traits that exhibit low heritability.
7. DNA markers permit QTL mapping.
8. Molecular markers are useful in indirect selection.
9. MAS can be useful for traits that are expressed late in development.
10. MAS require high technical skill than phenotypic selection.

E. FILL IN THE BLANKS

1. Marker characters refer to those traits that can be easily ________.
2. Marker assisted selection is also known as ____________ selection (MAS).
3. Marker assisted selection is based on __________.
4. MAS can be used as an aid to traditional __________ selection.
5. The results of MAS are consistent due to little __________ variation
6. MAS can be useful for traits that exhibit low ______________.
7. DNA markers permit _________ mapping.
8. Molecular markers are useful in _________ selection.
9. MAS can be useful for traits that are expressed ________in development.
10. MAS require high __________ than phenotypic selection.

F. ONE WORD OR SHORT ANSWER

1. Any genetic element (locus, allele, DNA sequence or chromosome feature) which can be readily detected by phenotype, cytological or molecular techniques.
2. Selection for specific alleles (which affect a trait of interest) using genetic markers.
3. A gene or other fragment of DNA whose location in the genome is known.
4. In plant breeding, markers that are related to variation in shape, size, color and surface of various plant parts.

5. Markers that are related to variation in proteins and amino acid banding pattern.
6. Markers that are related to variation in chromosome number, shape and size.
7. Improvement of crop plants for various economic characters through indirect selection for linked molecular markers.
8. The process of determining the genotype of an individual with a biological assay
9. Any observable characteristic of an organism, such as its morphology, development, biochemical or physiological properties, or behavior.
10. Presence of multiple alleles of a gene within a population, usually expressing different phenotypes
11. A fixed position on a chromosome, such as the position of a gene or a biomarker (genetic marker).
12. Influence of a single gene on multiple phenotypic traits.
13. The interaction between genes of different loci.

ANSWERS

C. MULTIPLE CHOICE QUESTIONS

Sl.No.	*Answer*	*Sl.No.*	*Answer*
1	(d) All of these	2	(d) All of these
3	(b) Molecular	4	(d) All of these
5	(d) All of these	6	(d) All of these
7	(d) All of these	8	(a) Morphological
9	(c) Biochemical	10	(d) Cytological

D. TRUE OR FALSE

1 to 10 all are true.

E. FILL IN THE BLANKS

Sl.No.	*Answer*	*Sl.No.*	*Answer*
1	Identified	2	Marker aided
3	Genetic markers	4	Phenotypic
5	Environmental	6	Heritability
7	QTL	8	Indirect
9	Late	10	Technical skill

F. ONE WORD OR SHORT ANSWER

Sl.No.	*Answer*	*Sl.No.*	*Answer*
1	Marker	2	Marker assisted selection
3	DNA Marker	4	Morphological Marker
5	Biochemical Marker	6	Cytological Marker
7	Molecular Breeding	8	Genotyping
9	Phenotype	10	Polymorphism
11	Locus	12	Pleiotropy
13	Epistasis		

Section III

Seed Production and Crop Breeding

Chapter 12

Varietal Seed Production

A. DEFINITIONS

Grow out Test: A test which is conducted to assess the genetic purity of a variety during seed production.

Isolation Distance: The separation of the field of a variety from that of another variety of the same crop by a minimum prescribed distance to avoid contamination.

Rogueing: The process of removal of the off types plants from the field of an improved variety to avoid contamination.

Seed Certification: A legal system which ensures availability of high quality seed of improved variety to the farmers.

Volunteer plants: The plants which grow in the field from previous crops.

B. BACKGROUND INFORMATION

1. Seed quality is determined by a number of genetic and physiological characteristics.
2. The genetic factors that can influence quality include genetic make-up, seed size and bulk density.
3. Deterioration in seed quality may begin at any point in the plant's development stage from fertilization onward.
4. Seed quality depends upon the physical conditions that the mother plant is exposed during its growth stages, as well as harvesting, processing, storage and planting.
5. Temperature, nutrients and other environmental factors also affect seed development and influence seed quality.
6. High quality seeds are the result of good production practices.

7. The production of good quality seed is based on certain genetic principles as well as agronomic principles.
8. Genetic principles include (i) control of seed source, (ii) isolation distance, (iii) rouging of off types, (iv) seed certification and (v) grow out test.
9. For quality seed production, the seed of improved category should be used and obtained from an approved source.
10. Four classes of seeds are generally recognized in seed certification chain namely breeder seed, foundation registered and certified. These classes are recognized by Association of official seed certifying agencies [AOSCA].
11. Isolation is essential to prevent pollination from unwanted pollen in the case of cross pollinated and often cross pollinated species to avoid mechanical mixture and chance cross pollination in self-pollinated crops.
12. The isolation distance varies from 3m in self-pollinated crops like rice, wheat to 200m in maize *etc.*
13. The presence of off type [phenotypically different] plants is another source of genetic contamination. Their continued presence would certainly deteriorate the genetic purity of the variety.
14. The objective of seed certification is to make available crop seeds, tubers, bulbs, *etc.* of good quality to the farmers. Seed certification work is carried out by well experienced and qualified persons of seed certification agency.
15. Certification requires two, three or four inspections depending on the crop species.
16. Grow out test is used to observe the genetic purity of the crop. Varieties being grown for seed production should periodically be tested for genetic purity by grow – out – test, to make sure that, seed being maintained in their true form.
17. Agronomic principles include (i) selection of field, (ii) cultural practices, (iii) plant protection, (iv) weed control, (v) harvesting and threshing, (vi) cleaning, grading and testing and bagging.
18. Regions of moderate rainfall and humidity are more suited for seed production. Most crops require a dry sunny period and moderate temperature for flowering and pollination.
19. Excessive dew and rainfall lead to poor seed set, high pest and disease incidence, heavy seed loss and lodging at harvest. High temperature results in pollen desiccation at flowering, premature flowering and poor quality seeds.
20. Selection of field for seed plot should meet certain requirements. The field should have good soil texture and fertility. It should be free from volunteer plants and weed plants. It should be free from soil borne diseases and insect pests. The plot must not have the same crop in the previous season also. The field should be well leveled with proper isolation. The field should be free from destruction of potential weeds.

21. Standard agronomic practices such as sowing time, spacing, fertilizer dose, hoeing and weeding have to be strictly followed to raise a good seed crop.
22. Proper crop rotation has to be adopted to avoid contamination from volunteers of the same crop. The seed production should be taken up under irrigated conditions to ensure high yield and good quality.
23. Incidence of some diseases may lead to the rejection of seed by the certifying agency as unfit for use. For example, maximum permissible limit for head smut in Sorghum is 1/1000. Hence, proper plant protection measures should be adopted to raise a healthy crop.
24. Effective weed control is a must for good seed production. Weeds reduce crop yield and their seeds contaminate the crop seed. It acts as host for number of pathogens. Some weeds are classified as noxious or objectionable weeds. Their seeds are similar to crop seeds and sometimes it is very difficult to separate them from seed of the concerned crops.
25. Considerable care should be taken to prevent mechanical mixture from other crops or weed seeds during harvesting and threshing. Threshing floor should be clean, while threshing. There should not be any damage to seeds.
26. Seed processing includes drying, cleaning, grading, treatment with chemicals, bagging and labeling. The seed should be dried to appropriate moisture level to facilitate processing to prevent losses in germination and to reduce the chances of insect attack during storage.
27. The seeds can be sun dried or can be dried by hot air.

C. MULTIPLE CHOICE QUESTIONS

1. Genetic principles of seed production include:

(a) Seed source (b) Isolation distance
(c) Roguing (d) All of these

2. Agronomic principles of seed production include:

(a) Selection of field (b) Cultural practices
(c) Plant protection (d) All of these

3. Agronomic principles of seed production include:

(a) Weed control (b) Harvesting and threshing
(c) Transport (d) All of these

4. Agronomic principles of seed production include:

(a) Cleaning (b) Grading
(c) Bagging (d) All of these

5. **What is isolation distance for foundation seed production of wheat?**
   (a) 3 m (b) 5m
   (c) 7m (d) 10m
6. **What is isolation distance for certified seed production of cotton?**
   (a) 20m (b) 30m
   (c) 40m (d) 50m
7. **What is isolation distance for foundation seed production of black gram?**
   (a) 5m (b) 10m
   (c) 20m (d) 30m

D. TRUE OR FALSE

1. Use of quality seed helps in quality seed production.
2. Effective weed control is a must for quality seed production.
3. Isolation is essential to prevent pollination from unwanted pollen.
4. Roguing prevents contamination with other types.
5. Proper crop rotation prevents contamination from volunteers of the same crop.
6. Adoption of proper plant protection measures is essential to raise a healthy crop.
7. Effective weed control is a must for quality seed production.
8. The seed drying is essential to reduce the chances of insect attack during storage.
9. The field should be free from objectionable weeds.
10. Proper care during harvesting, transport and threshing prevents mechanical mixture.

E. FILL IN THE BLANKS

1. Use of quality seed helps in quality ———— production.
2. Effective weed control is a must for ———— seed production.
3. Isolation is essential to prevent pollination from unwanted ————.
4. Rogueing prevents contamination with ———— types.
5. Proper crop rotation prevents contamination from ———— of the same crop.
6. Adoption of proper plant protection measures is essential to raise a ———— crop.
7. Effective weed control is a must for ———— seed production.

8. The seed drying is essential to reduce the chances of insect attack during ________.
9. The field should be free from ________ weeds.
10. Proper care during harvesting, transport and threshing prevents ________.

F. ONE WORD OR SHORT ANSWER

1. The separation of the field of a variety from that of another variety of the same crop by a minimum prescribed distance to avoid contamination.
2. The process of removal of the off types plants from the field of an improved variety to avoid contamination.
3. A legal system which ensures availability of high quality seed of improved variety to the farmers.
4. A test which is conducted to assess the genetic purity of a variety during seed production.
5. The plants which grow in the field from previous crops.

ANSWERS

C. MULTIPLE CHOICE QUESTIONS

Sl.No.	Answer	Sl.No.	Answer
1	(d) All of these	5	(a) 3 m
2	(d) All of these	6	(b) 30m
3	(d) All of these	7	(c) 20m
4	(d) All of these		

D. TRUE OR FALSE

1 to 10 all are true.

E. FILL IN THE BLANKS

Sl.No.	Answer	Sl.No.	Answer
1	Seed	6	Healthy
2	Quality	7	Quality
3	Pollen	8	Storage
4	Other	9	Objectionable
5	Volunteers	10	Mechanical mixture

F. ONE WORD OR SHORT ANSWER

Sl.No.	Answer	Sl.No.	Answer
1	Isolation Distance	4	Grow out Test
2	Roguing	5	Volunteer plants
3	Seed Certification		

Chapter 13

Hybrid Seed Production

A. DEFINITIONS

Breeder Seed: The progeny of nucleus seed or breeder seed produced under the strict supervision of original or sponsoring plant breeder.

Certified Seed: The progeny of either foundation or registered or certified seed. The certification is done by the State Seed Certification Agency.

Foundation Seed: The progeny of breeder seed produced by National Seeds Corporation.

Genetic Purity: The absence of seeds of other variety of the same crop as well as of other crops.

Germination: Emergence of normal seedlings from the seeds under ideal conditions of light, temperature, moisture, oxygen and nutrients.

Inert matter: Non- living materials such as sand, pebbles, soil particles, straw, *etc.*

Isolation: Separation of the field of a variety from that of another variety of the same crop to prescribed standard distance to avoid contamination.

Nucleus Seed: The initial seed of an improved variety, limited in quantity and produced by originating plant breeder.

Physical Purity: Freedom of seed from inert matter and defective seeds.

Registered Seed: The progeny of either foundation or registered seed. In India, this category is generally omitted and certified seed is produced directly from foundation seed.

Roguing: The process of removal of off type (phenotypically different) plants from the field of an improved variety to avoid contamination.

Seed Certification: A legal system which ensures production of high quality seed in terms of genetic purity and germination.

Seed testing: The process of evaluation of seeds in terms of purity and germination.

Varietal Deterioration: Permanent reduction either in the genetic or agronomic value of a released variety.

Variety: A genotype released for commercial cultivation either by State Variety Release Committee or Central Variety Release Committee.

B. BACKGROUND INFORMATION

1. The progeny of a cross between genetically dissimilar parents is called hybrid.
2. Hybrid variety refers to those F_1 populations which are used for commercial cultivation.
3. Based on involvement of parents in the cross, hybrids may be single cross hybrid, three-way cross hybrid and double cross hybrid.
4. Based on cultivation season, hybrids are classified in two groups, *viz.* kharif season hybrids and rabi season hybrids.
5. Kharif season hybrids are developed in maize, Sorghum, pearl millet, pigeon pea, rice and castor.
6. Rabi season hybrids are developed in rapeseed and mustard, safflower and sunflower.
7. Hybrid seed production program differs from crop to crop. However, there are some steps which are common in hybrid seed production of different crops. These steps include: (i) planting ratio, (ii) isolation distance, (iii) rogueing, and (iv) field inspections.
8. Planting ratio of female and male parents for hybrid seed production differs from crop to crop as presented in the following Table.

Table 13-1: Planting Ratio of Female and Male Parents for Hybrid Seed Production in different Field Crops

Sl.No.	*Name of Crop*	*Female :Male*	*Isolation Distance m*	*Planted in*
A	**Kharif Hybrids**			
1	Maize	6: 2	200	Adjacent rows
2	Sorghum	4 : 2	100	Adjacent rows
3	Pearl Millet	4 : 2	200	Adjacent rows
4	Pigeon Pea	6 : 1	200	Adjacent rows and adjacent blocks
5	Rice	8 : 2	3	Adjacent rows of A and R lines
6	Cotton	3 : 1	30	Adjacent rows
7	Castor	6 : 2	100	Adjacent rows

Sl.No.	Name of Crop	Female :Male	Isolation Distance m	Planted in
B	**Rabi Hybrids**			
8	Rapeseed and Mustard	4 : 2 or 2 :1	200	Adjacent rows of A and R lines
9	Safflower	6 : 2	200	Adjacent rows and adjacent blocks
10	Sunflower	6 : 2	200	Adjacent rows

9. Successful production of hybrid seed depend on several factors such as (i) purity of parental lines, (ii) stability of male sterile line, (iii) restoration capacity of R line, (iv) resistance level of parental lines, (v) Yield level of parental lines, (vi) nicking of female and male lines, and (vii) presence of marker in male sterile line.

C. MULTIPLE CHOICE QUESTIONS

2. How much isolation distance is required for hybrid seed production of maize?

(a) 200 m (b) 100 m

(c) 400 m (d) None of these

3. How much isolation distance is required for hybrid seed production of Sorghum?

(a) 200 m (b) 100 m

(c) 400 m (d) None of these

4. How much isolation distance is required for hybrid seed production of Pearl millet?

(a) 200 m (b) 100 m

(c) 400 m (d) None of these

5. How much isolation distance is required for hybrid seed production of Pigeon pea?

(a) 200 m (b) 100 m

(c) 400 m (d) None of these

6. How much isolation distance is required for hybrid seed production of Cotton?

(a) 30 m (b) 50 m

(c) 100 m (d) None of these

7. How much isolation distance is required for hybrid seed production of Rapeseed and mustard?

(a) 200 m (b) 100 m

(c) 400 m (d) None of these

8. **How much isolation distance is required for hybrid seed production of Sunflower?**
   (a) 200 m (b) 100 m
   (c) 400 m (d) None of these

9. **How much isolation distance is required for hybrid seed production of Rice?**
   (a) 3 m (b) 10 m
   (c) 40 m (d) None of these

10. **How much ratio of female and male parents is required for hybrid seed production of Maize?**
    (a) 6:2 (b) 4:2
    (c) 3:1 (d) None of these

11. **How much ratio of female and male parents is required for hybrid seed production of Sorghum?**
    (a) 6:2 (b) 4:2
    (c) 3:1 (d) None of these

12. **How much ratio of female and male parents is required for hybrid seed production of Pigeon pea?**
    (a) 6:2 (b) 6:1
    (c) 3:1 (d) None of these

D. TRUE OR FALSE

1. An isolation distance of 200m is required for hybrid seed production of maize.
2. An isolation distance of 100m is required for hybrid seed production of Sorghum.
3. An isolation distance of 200m is required for hybrid seed production of Pearl millet.
4. An isolation distance of 200m is required for hybrid seed production of Pigeon pea.
5. An isolation distance of 30m is required for hybrid seed production of Cotton.
6. An isolation distance of 200m is required for hybrid seed production of Rapeseed and Mustard.
7. The ratio of 6:2 female and male parents is required for hybrid seed production of Maize.
8. The ratio of 6:1 female and male parents is required for hybrid seed production of Pigeon Pea.

9. The ratio of 4:2 female and male parents is required for hybrid seed production of Sorghum.
10. The ratio of 6:2 female and male parents is required for hybrid seed production of Pearl millet.
11. The ratio of 3:1 female and male parents is required for hybrid seed production of Cotton.

E. FILL IN THE BLANKS

1. An isolation distance of ______ is required for hybrid seed production of maize.
2. An isolation distance of ______ is required for hybrid seed production of Sorghum.
3. An isolation distance of ______)is required for hybrid seed production of Pearl millet.
4. An isolation distance of ______ is required for hybrid seed production of Pigeon pea.
5. An isolation distance of _____ is required for hybrid seed production of Cotton.
6. An isolation distance of ______ is required for hybrid seed production of Rapeseed and Mustard.
7. The ratio of _____ female and male parents is required for hybrid seed production of Maize.
8. The ratio of _____ female and male parents is required for hybrid seed production of Pigeon Pea.
9. The ratio of _____ female and male parents is required for hybrid seed production of Sorghum.
10. The ratio of _____ female and male parents is required for hybrid seed production of Pearl millet.
11. The ratio of _____ female and male parents is required for hybrid seed production of Cotton.

F. ONE WORD OR SHORT ANSWER

1. A genotype released for commercial cultivation either by State Variety Release Committee or Central Variety Release Committee.
2. The initial seed of an improved variety, limited in quantity and produced by originating plant breeder.
3. The progeny of nucleus seed or breeder seed produced under the strict supervision of original or sponsoring plant breeder.
4. The progeny of breeder seed produced by National Seeds Corporation.

5. The progeny of either foundation or registered or certified seed. The certification is done by the State Seed Certification Agency.
6. The progeny of either foundation or registered seed. In India, this category is generally omitted and certified seed is produced directly from foundation seed.
7. A legal system which ensures production of high quality seed in terms of genetic purity and germination.
8. The absence of seeds of other variety of the same crop as well as of other crops.
9. Freedom of seed from inert matter and defective seeds.
10. Non- living materials such as sand, pebbles, soil particles, straw, *etc.*
11. Emergence of normal seedlings from the seeds under ideal conditions of light, temperature, moisture, oxygen and nutrients.
12. Separation of the field of a variety from that of another variety of the same crop to prescribed standard distance to avoid contamination.
13. The process of removal of off type (phenotypically different) plants from the field of an improved variety to avoid contamination.
14. Permanent reduction either in the genetic or agronomic value of a released variety.
15. The process of evaluation of seeds in terms of purity and germination.

ANSWERS

C. MULTIPLE CHOICE QUESTIONS

Sl.No.	Answer	Sl.No.	Answer
1	(a) 200 m	2	(b) 100 m
3	(a) 200 m	4	(a) 200 m
5	(a) 30 m	6	(a) 200 m
7	(a) 200 m	8	(a) 3 m
9	(a) 6:2	10	(b) 4:2
11	(b) 6:1		

D. TRUE OR FALSE

1 to 11 all are true

E. FILL IN THE BLANKS

Sl.No.	Answer	Sl.No.	Answer
1	200m	2	100m
3	200m	4	200m
5	30m	6	200m
7	6:2	8	6:1
9	4:2	10	6:2
11	3:1		

F. ONE WORD OR SHORT ANSWER

Sl.No.	Answer	Sl.No.	Answer
1	Variety	2	Nucleus Seed
3	Breeder Seed	4	Foundation Seed
5	Certified Seed	6	Registered Seed
7	Seed Certification	8	Genetic Purity
9	Physical Purity	10	Inert matter
11	Germination	12	Isolation
13	Roguing	14	Varietal Deterioration
15	Seed testing		

Chapter 14

Wheat and Rice Breeding

A. DEFINITIONS

Cereal Crops: The starchy grain of various grasses such as wheat, oats, or corn, which are used as foods refer to cereal crops.

Crop Plants: Domesticated and commercially cultivated plants or agricultural produce, such as grain, vegetables, or fruit, considered as a group are called crop plants: Wheat is a common crop.

Ear: Wheat and barley inflorescence.

Field Crop: A crop (other than fruits or vegetables) that is grown for agricultural purposes is called field crop. Cotton, pulses, sugarcane wheat, mustard, maize are examples of field crops"

Husk: In maize, leafy structures covering the ear.

Panicle: Rice inflorescence.

Silk: In maize, long stigmas which emerge from upper end of cob.

Tassel: In maize, the male inflorescence.

B. BACKGROUND INFORMATION

1. There are two cultivated species of wheat, *viz.* common wheat (*Triticum aestivum*) and durum wheat (*T. turgidum* L.).
2. Common wheat is hexaploid ($2n = 6x = 42$), whereas durum wheat is tetraploid ($2n = 4x = 28$). The former is more widely adapted than the latter. Common wheat is used for bread, cakes, noodles, cookies, chapati *etc.*, whereas the durum wheat is used mainly for macaroni and some flat bread.
3. There is onlyone International wheat breeding center, *i.e.* International maize and Wheat Improvement Centre (CIMMYT), Mexico. International

multi-locational testing is carried out by CIMMYT for identification and release of varieties for different countries. The global gene poll of wheat is also maintained at this centre.

4. Indian Institute of Wheat and Barley Research, Karnal. The new varieties are released through coordinated project after multi-location testing for 3-5 years.
5. In Wheat cross-pollination is less than 1 per cent. Flowers open generally after pollination is over (chasmogamy). This chamogampus promotes autogamy.
6. Some high yielding varieties of common wheat for different situations are listed below:
7. **Irrigated timely sown**: HD 2329, HD 2009, HD 2428, PBW 154, CPAN 3004 (Punjab, Haryana, Western
8. U.P., Rajasthan)
9. **Irrigated late sown**: HD 2285, HD 2270, PBW 226, PBW 138 (Above 4 states)
10. **Rain-fed timely sown**: C 306, WL 410, IWP 72, PBW 65, PBW 175, WL 2265 (Above States)
11. **Salt affected soils:** WH 157, KRL 1-4
12. **Hilly Regions:** CPAN 1796, HD 208, HD 2380, HS 240, UP 1109 (U.P., J&K and H.P.) Varieties of durum wheat are PBW 34, PBW 215 and Raj 1555 for irrigated timely sown areas and JK 12 for rainfed timely sown areas.
13. There are two cultivated species of rice, *viz.* Asian rice (*Oryza sativa* L.) and African rice (*Oryza glaberrima*). Both the species are diploid ($2n = 24$). In Rice cross-pollination is less than 1 per cent.
14. Asian rice is the predominant species which has spread to different parts of the world. The African rice is still confined to tropical West Africa.
15. There are three races of Asian rice, *viz.* Indica, Japonica and Javanica. Indica is grown in tropical climate, Japonica in the temperate region and the race Javanica is intermediate between Indica and Japonica.
16. Rice has originated in Asia as well as in Africa. The Asian rice [Oriza sativa] has originated from perennial Asian wild species [*Oryza rupipogan*] and the African cultivated rice [*Oryza glaberrema*] is believed to have originated in West Africa from wild diploid species [*Oryza breviligulata*]
17. There is only one International rice breeding center, *i.e.* International Rice Research Institute, Philippines. International multi-locational testing is carried out by IRRI for identification and release of varieties for different countries. The global gene poll of rice is also maintained at this centre.
18. In India, rice breeding work is carried out by two main centers, *viz.* National Rice Research Institute, Cuttack, Odisha, Indian Institute of Rice Research, Hyderabad, Telangana.

19. Important rice varieties developed by IRRI include, IR 8, IR 20, IR 22, IR 24, IR 26, IR 28, IR 29, IR 30, IR 32, IR 34 and IR 36. Out of these, IR 8, IR 20 and IR 36 have been released for India. In China, hybrid rice using cytoplasmic genetic male sterility system has been developed for commercial cultivation. Hybrid rice has also been developed in India by Indian Institute of Rice Research, Hyderabad.
20. In India, more than 100 improved varieties of rice for different states have been released by CRRI, DRR and State Agricultural Universities. The bacterial blight resistant varieties developed by IRRI and salinity resistant, scented and high yielding varieties developed for cultivation in India are listed below:
21. **Bacterial blight resistant varieties:** IR 20, IR 22, IR 26, IR 28, IR 29, IR 30, IR 32 and IR 34.
22. **Salinity resistant varieties**: Usar Dhan 1 (U.P.), PVR 1 (Tamil Nadu), MCM 1, MCM 2 (A.P.) Bhurarata, Kalarata (Maharashtra), Korekagga, Bilikagga (Karnataka), Damodar, Getu, Dosal, Nanobokhra, Nonasail, Rupsail (West Bengal), Pokkai and Chotupokkali (Kerala).
23. **Scented varieties**: Type 3, N 12, N 10 (B), Kasturi, Pusa, Basmati.
24. **High yielding varieties:** IR 8, IR 20, IR 36, Vikas, Savitri, Anamika, Mansarovar *etc.*
25. Norin 10 in wheat and Dee-Geo-Woo-gen in rice are sources of dwarfing gene.

C. MULTIPLE CHOICE QUESTIONS

1. IR 20 is the variety of which crop?

(a) Wheat (b) Barley
(c) Rice (d) None of these

2. In which crop Norin 10 is a source of dwarfing gene?

(a) Wheat (b) Barley
(c) Rice (d) None of these

3. In which crop Dee-Geo-Woo-Gen is a source of dwarfing gene?

(a) Wheat (b) Barley
(c) Rice (d) None of these

4. Where is IRRI located?

(a) Mexico (b) Philippines
(c) Cuttock (d) None of these

5. Where is NRRI located?

(a) Hyderabad (b) Ludhiana
(c) Cuttack (d) None of these

6. **Where is IIRR located?**
   (a) Hyderabad (b) Ludhiana
   (c) Cuttack (d) None of these
7. **Where is CIMMYT located?**
   (a) Mexico (b) Philippines
   (c) Cuttack (d) None of these
8. **Where is IIWBR located?**
   (a) Hyderabad (b) Karnal
   (c) Cuttack (d) None of these
9. **How much is cross-pollination in Wheat?**
   (a) 5 per cent (b) Less than 1 per cent
   (c) 2 per cent (d) None of these
10. **How much is cross-pollination in Rice?**
   (a) 5 per cent (b) Less than 1 per cent
   (c) 6 per cent (d) None of these

D. TRUE OR FALSE

1. Bread wheat is also known as soft wheat.
2. Durum wheat is also known as hard wheat.
3. Indian Institute of Wheat and Barley Research is located at Karnal.
4. International maize and Wheat Improvement Centre (CIMMYT) is located in Mexico.
5. National Rice Research Institute is located at Cuttack, Odisha.
6. Indian Institute of Rice Research is located at Hyderabad, Telangana.
7. International Rice Research Institute is located in Philippines.
8. Norin 10 is the source of dwarfing gene in wheat.
9. Dee-Geo-Woo-Gen is the source of dwarfing gene in Rice.
10. The *Oriza sativa* has originated from perennial Asian wild species *Oryza rupipogan.*
11. African cultivated rice [*Oriza glaberrema*] has originated in West Africa from wild diploid species [*Oryza breviligulata*].
12. In Wheat cross-pollination is less than 1 per cent.
13. In Rice cross-pollination is less than 1 per cent.
14. In Rice, the inflorescence is called panicle.

E. FILL IN THE BLANKS

1. Bread wheat is also known as _____ wheat.
2. Durum wheat is also known as _____ wheat.
3. Indian Institute of Wheat and Barley Research is located at ______.
4. International maize and Wheat Improvement Centre (CIMMYT) is located in ______.
5. National Rice Research Institute is located at _______.
6. Indian Institute of Rice Research is located at _____________.
7. International Rice Research Institute is located in ________.
8. Norin 10 is the source of dwarfing gene in _______.
9. Dee-Geo-Woo-Gen is the source of dwarfing gene in ______.
10. The Oriza sativa has originated from perennial Asian wild species _______.
11. African cultivated rice [Oriza glaberrema] has originated in West Africa from wild diploid species ________.
12. In Wheat cross-pollination is less than ______.
13. In Rice cross-pollination is less than _______.
14. In Rice, the inflorescence is called _______.

F. ONE WORD OR SHORT ANSWER

1. Domesticated and commercially cultivated plants or agricultural produce, such as grain, vegetables, or fruit, considered as a group are called crop plants: Wheat is a common crop.
2. A crop (other than fruits or vegetables) that is grown for agricultural purposes is called field crop. Cotton, pulses, sugarcane wheat, mustard, maize are examples of field crops"
3. The starchy grains of various grasses such as wheat, oats, or corn, which are used as food refer to cereal crops.
4. Wheat and barley inflorescence.
5. Rice inflorescence.

ANSWERS

C. MULTIPLE CHOICE QUESTIONS

Sl.No.	Answer	Sl.No.	Answer
1	(c) Rice	2	(a) Wheat
3	(c) Rice	4	(b) Philippines
5	(c) Cuttock	6	(a) Hyderabad
7	(a) Mexico	8	(b) Karnal
9	(b) Less than 1%	10	(b) Less than 1%

D. TRUE OR FALSE

1 to 14 all are true.

E. FILL IN THE BLANKS

Sl.No.	Answer	Sl.No.	Answer
1	Soft	2	Hard
3	Karnal	4	Mexico
5	Cuttack	6	Hyderabad
7	Philippines	8	Wheat
9	Rice	10	Oryza rupipogan
11	Oryza breviligulata	12	1 per cent
13	1 per cent	14	Panicle

F. ONE WORD OR SHORT ANSWER

Sl.No.	Answer	Sl.No.	Answer
1	Crop Plants	2	Field Crop
3	Cereal Crops	4	Ear
5	Panicle		

Chapter 15

Maize, Sorghum and Pearl Millet Breeding

A. DEFINITIONS

Cereal Crops: The starchy grain of various grasses such as wheat, oats, or corn, which are used as food refer to cereal crops.

Crop Plants: Domesticated and commercially cultivated plants or agricultural produce, such as grain, vegetables, or fruit, considered as a group are called crop plants: Wheat is a common crop.

Ear: Wheat and barley inflorescence.

Field Crop: A crop (other than fruits or vegetables) that is grown for agricultural purposes is called field crop. Cotton, pulses, sugarcane wheat, mustard, maize are examples of field crops"

Husk: In maize, leafy structures covering the ear.

Panicle: Rice inflorescence.

Silk: In maize, long stigmas which emerge from upper end of cob.

Tassel: In maize, the male inflorescence.

B. BACKGROUND INFORMATION

1. Maize is a monoecious plant. Male and female flowers are separate on the same plant. Male inflorescence is called tassel and female flower is known as silk. Tassel is present on the top and silk in the middle portion of the plant.
2. In maize, an isolation distance of 400 m and 200 m is required for the production of foundation seed of open pollinated varieties.

3. International maize and Wheat Improvement Centre (CIMMYT), Mexico. International multi-locational testing is carried out by CIMMYT for identification and release of varieties for different countries.
4. Indian Institute of Maize Research, is located at Ludhiana. The new varieties are released through coordinated project after multi-location testing for 3-5 years.
5. Vivekanand Parvatiya Krishi Anusandhan Shala, Almora, develops maize varieties especially for hill regions.
6. State Agricultural Universities located in maize growing regions also develop maize varieties.
7. **Composite:** Jawahar, Vijay, Kisan, Amber, Sona, Vikram, Protina, Rattan, Shakti, Vikas, Navjot, Ageti 76, Kanchan, Diara 3, D 765, MCU 508, Kiran, Surya, Tarun, Arun, Renuka, *etc.*
8. **Hybrids:** Ganga Safed, Ganga 5, Ganga 9, Deccan 103, Sartaj Makka 1, VL 42 *etc.* Hybrids Him 123 and Him 128 have been released for the hilly regions of UP, HP, Assam and Sikkim.
9. Both genetic and cytoplasmic genetic male sterility systems are available in maize. Usually, CGMS system is used for the development of single, three-way and double cross hybrids in maize.
10. It is generally accepted that Ethopia is the centre of origin of cultivated *Sorghum*. From there it spread to other parts of the world. The diploid wild species *S. arundinaceum* ($2n = 20$) is considered to be the likely ancestor of cultivated *Sorghum*. This wild species is still found growing in Ethopia and its adjacent regions.
11. At International level Sorghum improvement work is carried out by International Crop Research Institute for Semi-arid Tropics (ICRISAT), Hyderabad (India). New varieties are developed after multi-location testing in different countries.
12. At National level Sorghum improvement work is carried out by Indian Institute of Millet Research, Hyderabad.
13. Sorghum is a self-pollinated crop. Natural cross pollination varies from 5 to 30 per cent depending upon genotype and environmental conditions. Sorghum pollen is very light and cross pollination occurs by wind.
14. In Sorghum, an isolation distance of 200 m. and 100 m. is essential for the production of foundation and certified seed.
15. In India, several high yielding varieties and hybrids have been developed for different states. Hybrids are developed by using cytoplasmic genic male sterility. Kafir 60 is the source of male sterility. High lysine Ethiopian mutants (IS 11167 and 11758) are used for quality improvement. Some important high yielding varieties and hybrids of grain Sorghum are listed as follows:

16. **Hybrids:** CSH 1, CSH 2, CSH 3, CSH 4, CSH 5, CSH 6, CSH 7, CSH 8, CSH 9, CSH 10 and CSH 11 for Kharif and , CSH 7, CSH 8, CSH 12, and CSH 13 for Rabi season.
17. **Varieties:** CSV 1, CSV 2, CSV 3, CSV 4, CSV 5, CSV 6, CSV 7, CSV 8, CSV 9, CSV 10 and CSV 11, CSV 12, CSV 13, M 35-1 and Swati for Kharif and CSV 7 and CSV 8, for Rabi season.
18. Pearl millet is a cross-pollinated crop. In this crop natural cross pollination is more than 80 per cent. Pollen is very light and cross pollination occurs by wind.
19. In Pearl millet an isolation distance of 400 m. and 200 m is essential for the production of foundation and certified seed.
20. At International level pearl millet improvement work is carried out by International Crop Research Institute for Semi-arid Tropics (ICRISAT), Hyderabad (India).
21. At National level pearl millet improvement work is carried out by Central Arid-zone Research Institute, Jodhpur and some State Agricultural Universities.
22. Important varieties developed by Research Stations of Agricultural Universities and other organizations are listed below:
23. **Hybrids**: HHB 45, HHB 50, HHB 60, HHB 67 from Hisar; GHB 27, GHB 30, GHB 181 FROM Jamnagar [Gujarat]; Pusa 3 from IARI, New Delhi; PHB 17 from Punjab; Other hybrids include HBH 128, MBH 130, MH 143, MH 169, MH 179, MH 180, MH 82 MH 208, *etc*. All hybrids are CMS based. Tift 23 A and Ludhiana cytoplasm have been used as sources of CMS in pearl millet improvement.
24. **Varieties**: PSB 8, PSB 15, PSB 2, HC 4, RHR 1, PCB 15, Mukta, Mallikarjun, Pusa safed Bajra, *etc*.

C. MULTIPLE CHOICE QUESTIONS

1. How much isolation distance is required for production of foundation seed in Sorghum?

(a) 100 m (b) 200 m
(c) 300 m (d) None of these

2. How much isolation distance is required for production of certified seed in Sorghum?

(a) 100 m (b) 200 m
(c) 300 m (d) None of these

3. How much isolation distance is required for production of foundation seed in Pearl millet?

(a) 200 m (b) 400 m
(c) 800 m (d) None of these

4. How much isolation distance is required for production of Certified seed in Pearl millet?

(a) 100 m (b) 200 m

(c) 300 m (d) None of these

5. How much is cross-pollination in Maize?

(a) 80 per cent (b) More than 99 per cent

(c) 60 per cent (d) None of these

6. How much is cross-pollination in Pearl millet?

(a) >80 per cent (b) More than 99 per cent

(c) 60 per cent (d) None of these

7. Where is CAZRI located?

(a) Hyderabad (b) Ludhiana

(c) Jodhpur (d) None of these

8. Where is Indian Institute of Maize Research located?

(a) Hyderabad (b) Ludhiana

(c) Cuttock (d) None of these

9. Where is Indian Institute of Millets Research located?

(a) Hyderabad (b) Jodhpur

(c) Cuttock (d) None of these

10. In which crop Kafir 60 is a source of male sterility?

(a) Maize (b) Sorghum

(c) Pearl millet (d) None of these

D. TRUE OR FALSE

1. In Maize the photosynthetic pathway is C4.
2. The place of origin of maize is Mexico.
3. Maize is a cross-pollinated and seed propagated crop.
4. In maize, cross-pollination is more than 99 per cent.
5. In Maize pollination occurs by wind.
6. In maize male inflorescence is called tassel.
7. Non-tillering cereal is Maize.
8. In Sorghum, protogyny promotes cross pollination.
9. The cultivated species of pearl millet is. *Pennisetum americanum.*
10. The original home of pearl millet is Western Africa.

11. Indian Institute of Maize Research is located at Ludhiana.
12. Indian Institute of Millets Research is located at Hyderabad.

E. FILL IN THE BLANKS

1. In Maize the photosynthetic pathway is ______.
2. The place of origin of maize is _______.
3. Maize is a _____pollinated crop.
4. In maize, cross-pollination is more than ______.
5. In Maize pollination occurs by ______.
6. In maize male inflorescence is called ______.
7. Non-tillering cereal is ________.
8. In Sorghum, protogyny promotes _________.
9. The cultivated species of pearl millet *is*. ________.
10. The original home of pearl millet is _______.
11. Indian Institute of Maize Research is located at ________.
12. Indian Institute of Millets Research is located at _________.

F. ONE WORD OR SHORT ANSWER

1. Domesticated and commercially cultivated plants or agricultural produce, such as grain, vegetables, or fruits, are called crop plants: Wheat is a common crop.
2. A crop (other than fruits or vegetables) that is grown for agricultural purposes is called field crop. Cotton, pulses, sugarcane wheat, mustard, maize are examples of field crops"
3. The starchy grain of various grasses such as wheat, oats, or corn, which are used as food refer to cereal crops.
4. In maize, the male inflorescence.
5. In maize, long stigmas which emerge from upper end of cob.
6. In maize, leafy structures covering the ear.
7. Wheat and barley inflorescence.
8. Rice inflorescence.

ANSWERS

C. MULTIPLE CHOICE QUESTIONS

Sl.No.	Answer	Sl.No.	Answer
1	(b) 200 m	2	(a) 100 m
3	(b) 400 m	4	(b) 200 m
5	(b) More than 99%	6	(a) >80 %
7	(c) Jodhpur	8	(b) Ludhiana
9	(a) Hyderabad	10	(b) Sorghum

D. TRUE OR FALSE

1 to 12 all are true.

E. FILL IN THE BLANKS

Sl.No.	Answer	Sl.No.	Answer
1	C4	2	Mexico
3	Cross	4	99 per cent
5	Wind	6	Tassel
7	Maize	8	Cross pollination
9	*Pennisetum americanum*	10	Western Africa
11	Ludhiana	12	Hyderabad

F. ONE WORD OR SHORT ANSWER

Sl.No.	Answer	Sl.No.	Answer
1	Crop Plants	2	Field Crop
3	Cereal Crops	4	Tassel
5	Silk	6	Husk
7	Ear	8	Panicle

Chapter 16

Breeding Commercial Crops

A.DEFINITIONS

Cotton: Those species of the genus Gossypium which produce sinnable seed coat fibres.

Fibre Crops: Field crops which are grown for their fibres, which are traditionally used to make paper, cloth, or rope are called fibre crops. Such crops include cotton, Jute, sunnhamp, Kenef, *etc.*

Field Crop: A crop (other than fruits or vegetables) that is grown for agricultural purposes is called field crop. Cotton, pulses, sugarcane wheat, mustard, maize are examples of field crops.

Ginning: In cotton, the process of separating lint from the seed.

Monopodia: In cotton, vegetative branches.

Pulses: Those species of the family Leguminosae whose seeds are used for human consumption.

Square: In cotton, unopened floral bud.

Sympodia: In cotton, fruiting branches.

Test Weight: Weight of 100 seeds in gram.

Trash: In cotton, dried leaf bits or any other foreign matter in the lint.

B. BACKGROUND INFORMATION

1. Important commercial field crops include cotton, sugarcane, tobacco and potato.
2. Cotton is a fibre, oil and protein yielding crop. The lint is used in the textile mills, linters for various industrial uses, refined cotton seed oil is used for human consumption and the cake as fertilizer and livestock feed.

3. In the genus gossypiun about 50 species have been identified so far. Out of these four species, *viz.*, *Gossypium hirsutum*, *G. barbadense*, *G. arboreum* and *G. herbaceum* are cultivated and rest are wild or uncultivated.
4. The first two species are tetraploid [2n =52] and last two species are diploid [2n =26]. Tetraploid cottons are also known as New World cottons and diploid cottons are also known as Old World cottons or Asiatic cottons.
5. India is the only country where all the four cultivated species are commercially planted. The *G. hirsutum* is also known as upland cotton or American cotton. *G. barbadense* is also known as Egyptian cotton or Peruvian cotton or Tanguish cotton or Sea Island cotton.
6. It is believed that diploid species of cotton originated first and tetraplois species later on form a cross between two diploid species. The wild linted species *Gossypium africanum* is considered as the progenitor of all cultivated species of cotton.
7. Gossypium africanum reached undivided India from South Africa through travelers, traders and explorers and after differentiation gave birth to two diploid cultivated species, *viz. Gossypium arboretum Gossypium herbaceum*.
8. There are two species of tetraploid cultivated cotton, *viz. Gossypium hirsutum* and *G. barbadense*. The genome constitution of both these species is AD. It is believed that A genome of tetraploid cotton has come from Asian cultivated species and D genome from American lintless wild diploid species (Skovsted, 1934). Skovsted (1934) first pointed out that the New World tetraploid consisted of A and D sub-genomes. Later on origin of tetraploid cotton was explained by other workers.
9. Cotton is basically a self-pollinated crop. Cross pollination varies from nil to 60 per cent under different climatic conditions.
10. **In cotton,** an isolation distance of 50 m and 30 m is required for the foundation of and certified seed.
11. No international research centre for cotton breeding has been established so far. There is International Cotton Advisory Committee (ICAC) which advises on various matters of cotton research and marketing.
12. The important varieties and hybrids of cotton are listed below:
13. 1. Varieties
14. *G. hirsutum*. LH 900, LH 1556, F 846, F 1378, F 1861, HS 6, H 874, H 1098, HS 182, H 1117 RST 9, RS 875, 810 (North zone), Khandwa 3, Vikram, PKV 081, Rajat, LRA 5166, LRK 516 (Central zone), Abadita, Sharada, Sahana, MCU 12, MCU 5 VT, MCU 10, and LRA 5166. (South zone)
15. *G. arboreum*. LD 491, LD 694, RG 18, HD 107, HD 123, LD 327, RG 8 (North zone), AKH 4, AKA 8401, Maljari, AKA 5, AKA 7, PA 255, PA 402 (Central zone) K 9, K 10, K 11 (South zone).
16. (*a*) **Interspecific hybrids.** Varalaxmi, DCH 32, DHB 105, Sruthi TCHB 213,

NHB 12, HB 224 *etc.*

17. (*b*) **Intraspecific hybrids.** HH : H 4, H 6, H 8, H 10, NHH 44, JKHY 1, JKHY 2, PKV HY 2, PKVHY 3, PKVHY 4, PKVHY 5, Savita, CICR HH 1, Surya, DHH 11, Fateh, LHH 144, Dhanlaxmi, Maruvikas, Omshankar, CSHH 198, *etc.*
18. (*c*) **Desi hybrids.** DH 7, DH 9, DDH 2, Pha 46, LDH 11, AKDH 7, RAJDH 7, GCot MDH 11, AAH 1, CICR 2 *etc.*
19. (*d*) **Male Sterility based hybrids.** Suguna, PKVHY 3, PKVHY 4, PKVHY 4, AAH 1, ARDH 7, RAJDH7, G.Cot MDH 11, CICR 2.
20. (*e*) **Bt. Hybrids.** About 619 Bt. cotton hybrids have been released by different seed companies.
21. Sucrose is the main product of sugarcane. The other products are green tops, bagasse, molasses, filter mud, stubbles, and dry leaves. Green tops are used as cattle feed; bagasse for production of paper, rayon grade pulp, particle boards, fibre boards, as a fuel and also for production of alcohol; molasses for production of citric acid, butamol, acetone, rum, glycol, alcohol and acetic acid; filter mud for production of crude wax, fatty lipids, waxy lipids and refined wax; stubles for making compost and as fuel; and dry leaves as mulch, fuel and compost.
22. There are three cultivated species of sugarcane, *viz. Saccharum officinarum* ($2n = 80$), *S. sinense* ($2n = 118$) and *S. barberi* ($2n = 82\text{-}124$). The species *S. officinarum* is also known as noble cane. There are two wild species of sugarcane, *viz. S. spontaneum* and *S. robustum.*
23. It is believed that Saccharum officinarum has originated in New Guinea and S. barberi and S. sinensis in North India. Wild species S. robustum is considered to be the progenitor of cultivated species S. officinarum.
24. Sugarcane is a cross-pollinated crop and asexually propagated crop.
25. **In sugarcane,** an isolation distance of 50 m and 30 m is required for the foundation of and certified seed.
26. No International Research Institute has been established for sugarcane breeding so far. In India, sugarcane breeding work is carried out by Sugarcane Breeding Institute, Coimbatore and. Indian Institute of Sugarcane Research, Lucknow.
27. Several improved varieties of sugarcane suitable for different states have been released by Sugarcane Breeding Institute, Coimbatore and State Sugarcane Research Stations through Coordinated Project. Some currently cultivated sugarcane varieties are Co 1148, Co 1158, Co 64, Co 997, Co 6304, Co 419, Co 740, Bo 17 and Co Pant 84211. Some varieties with special characters are given below:

    Early varieties: Co 8336, Co 8337, Co 8338, Co 8339, Co 8340, Co 8341, (8-9 months) Co J64, and Co C 671

Drought Resistant: Co 285, Co 740, Co 997, Co 1148 Frost : Co 1148, N Co 310

Salinity Resistant: Co 453, Co 62125

Lodging Resistant: Co 6304, Co 7117, CoS 7918

Water logging Resistant : Co 1157, Co 975, Co 785, Bo 91, Bo 104, Bo 106, Bo 109, Bo 120, Cos 510, CoS 767, CoS 8016.

Top borer Resistant : COJ 67, Co 1158, Co 7224

Internode borer Resistant : COC 671, Co 975, Co 6806, Co 62175. **Red rot**: Co 7627, COJ 64, COR 8001; CoLK 8001, COH 7803, Bo 108.

28. The Central Andean region of South America is the original home of potato. Cultivated potato [Solanum tuberosum] is a tetraploid species [2n=48] which is believed to have originated from the diploid species S. stenotomum which is also found growing in the Andes region of South Peru and Bolivia.
29. Potato is a self-pollinated and vegetatively propagated crop.
30. International potato Centre (CIP), Peru serves as a world centre for genetic improvement of potato. The global collection of potato is also maintained at this centre. The main focus of this centre is on the improvement of potato for tropical and temperate climates of different countries. It has testing centres in different countries.
31. In India, potato breeding work is carried out by Central Potato Research Institute (CPRI), Shimla (Himachal Pradesh).
32. CPRI, Shimla has done remarkable work on potato improvement. Several high yielding varieties for various areas have been developed by this Institute. Some important varieties are Kufri Red, Kufri Sindhuri, Kufri Chamatkar, Kufri Kuber, Kufri Kundan, Kufri Alankar, Kufri Chandramukhi and Kufri Jyoti.

C. MULTIPLE CHOICE QUESTIONS

1. Where is Sugarcane breeding Institute located?

(a) Lucknow (b) Coimbatore

(c) Modipuram (d) None of these

2. Where is Indian Institute of Sugarcane Research located?

(a) Lucknow (b) Coimbatore

(c) Modipuram (d) None of these

3. Where is Central Potato Research Institute located?

(a) Shimla (b) Coimbatore

(c) Modipuram (d) None of these

4. Where is International Potato Center located?

(a) Shimla (b) Coimbatore

(c) Peru (d) None of these

5. Where is Central Institute for Cotton Research located?

(a) Nagpur (b) Coimbatore

(c) Ludhiana (d) None of these

6. Where is headquarters of International Cotton Advisory Committee located?

(a) New Delhi (b) Washington DC

(c) Tashkent (d) None of these

7. Where is Central Institute for Research on Cotton Technology located?

(a) Nagpur (b) Mumbai

(c) Ludhiana (d) None of these

8. Which of the following is related to Maize?

(a) Non-tillering (b) Monoecious nature

(c) C4 Pathway (d) All of these

9. How much isolation distance is required for production of foundation seed in sugarcane?

(a) 30 m (b) 50 m

(c) 100 m (d) None of these

10. How much isolation distance is required for production of foundation seed in Cotton?

(a) 30 m (b) 50 m

(c) 100 m (d) None of these

D. TRUE OR FALSE

1. The photosynthetic pathway in Sugarcane is C4.
2. Potato is called as king of vegetables.
3. Central Potato Research Institute (CPRI) is located at Shimla.
4. International potato Centre (CIP) is located in Peru.
5. Central Andean region of South America is the original home of potato.
6. Sugarcane Breeding Institute is located at Coimbatore.
7. Indian Institute of Sugarcane Research is located at Lucknow.
8. In sugarcane, an isolation distance required for production of foundation seed is 50 m.

9. International Cotton Advisory Committee (ICAC) is located at Washington, DC, USA.
10. Central Institute for Cotton Research (CICR) is located at Nagpur.
11. Central Institute for Research on Cotton Technology (CIRCOT) is located at Mumbai.

E. FILL IN THE BLANKS

1. The photosynthetic pathway in Sugarcane is ______.
2. Potato is called as _______of vegetables.
3. Central Potato Research Institute (CPRI) is located at ______.
4. International potato Centre (CIP) is located in ______.
5. The original home of potato is Andean region of ________
6. Sugarcane Breeding Institute is located at ______.
7. Indian Institute of Sugarcane Research is located at _______.
8. In sugarcane, an isolation distance required for production of foundation seed is ______.
9. International Cotton Advisory Committee (ICAC) is located at _______ USA.
10. Central Institute for Cotton Research (CICR) is located at _______.
11. Central Institute for Research on Cotton Technology (CIRCOT) is located at _______.

F. ONE WORD OR SHORT ANSWER

1. A crop (other than fruits or vegetables) that is grown for agricultural purposes is called field crop. Cotton, pulses, sugarcane wheat, mustard, maize are examples of field crops"
2. Field crops which are grown for their fibers, which are traditionally used to make paper, cloth, or rope are called fibre crops. Such crops include cotton, Jute, sunnhemp, Kenef, *etc.*
3. Weight of 100 seeds in gram.
4. Those species of the genus Gossypium which produces spinnable seed coat fibres.
5. In cotton, the process of separating lint from the seed.
6. In cotton, dried leaf bits or any other foreign matter in the lint.
7. In cotton, unopened floral bud.
8. In cotton, vegetative branches.
9. In cotton, fruiting branches.

ANSWERS

C. MULTIPLE CHOICE QUESTIONS

Sl.No.	Answer	Sl.No.	Answer
1	(b) Coimbatore	2	(a) Lucknow
3	(a) Shimla	4	(c) Peru
5	(a) Nagpur	6	(b) Washington DC
7	(b) Mumbai	8	(d) All of these
9	(b) 50 m	10	(b) 50 m

D. TRUE OR FALSE

1 to 11 all are true.

E. FILL IN THE BLANKS

Sl.No.	Answer	Sl.No.	Answer
1	C4	2	King
3	Shimla	4	Peru
5	South America	6	Coimbatore
7	Lucknow	8	50 m
9	Washington, DC,	10	Nagpur
11	Mumbai		

F. ONE WORD OR SHORT ANSWER

Sl.No.	Answer	Sl.No.	Answer
1	Field Crop	2	Fibre Crops
3	Test Weight	4	Cotton
5	Ginning	6	Trash
7	Square	8	Monopodia
9	Sympodia		

Chapter 17

Breeding Oilseed Crops

A. DEFINITIONS

Cereal Crops: The starchy grain of various grasses such as wheat, oats, or corn, which are used as food refer to cereal crops.

Cotton: Those species of the genus Gossypium which produce sinnable seed coat fibres.

Crop Plants: Domesticated and commercially cultivated plants or agricultural produce, such as grain, vegetables, or fruit, considered as a group are called crop plants: Wheat is a common crop.

Field Crop: A crop (other than fruits or vegetables) that is grown for agricultural purposes is called field crop. Cotton, pulses, sugarcane wheat, mustard, maize are examples of field crops"

Food Crops: Crops that can be used for food for humans. Examples: wheat, corn, fruits, vegetables are known as food crops.

Shell: In groundnut, the pericarp of the pod.

Test Weight: Weight of 100 seeds in gram.

B. BACKGROUND INFORMATION

1. There are several oil yielding crops such as rapeseed and mustard, ground nut, soybean, sunflower, *etc.*
2. In India, Rapeseed and mustard breeding work is carried out by National Research Centre for Mustard, Bharatpur (Rajasthan) and State Agricultural Universities of Rapeseed and Mustard growing states.
3. New varieties are released after multi-location testing in coordinated trials for 3-5 years. The coordinated project was located at HAU, Hisar. Now this has been shifted to Bharatpur. Several high yielding varieties

of rapeseed and mustard have been released in India for different states by above research centres.

4. Important National varieties of mustard are Kranti, RLM 198, Krishna and RLM 514. Pant Toria 303 is the national variety of rapeseed.
5. Soybean improvement work is carried out by various International and National crop research centers as follows:
6. **International Centers:** Following Internal crop research centers are involved in soybean improvement.
7. **AVRDC**: Asian Vegetable Research and Development Center [AVRDC], Taiwan. AVRDC deals with improvement of tomato, Chinese cabbage, sweet potato, mung bean and soybean for Asian region.
8. **IITA**: International Institute of Tropical Agriculture, Nigeria. IITA deals with improvement of cowpea, soybean, maize, rice, root tubers and plantation crops.
9. **National Centers:** In India, soybean improvement work is carried out by following organizations:
10. Directorate of Soybean Research, Indore, Madhya Pradesh. The new varieties are released through coordinated project after multi-location testing for 3-5 years.
11. State Agricultural Universities located in soybean growing regions.
12. **Practical Achievements:** In India, several improved varieties of soybean have been developed for different states. Eight varieties (Bragg, Clark 63, Davis, Hardee, Improved Pelicon, Jupiter, Lee and Monetta) have been released through direct introduction. Some high yielding varieties have been developed by Pantnagar, Jabalpur, Ludhiana and other Agricultural universities and also by IARI, New Delhi. These varieties include PK 262, PK 308, PK 327, PK 416, PK 472, Shilajeet and Pant Soybean 564 (Pantnagar); Durga, Gaurav, JS 75-46, JS 80-21, (*Jabalpur); SL 4 and SL 96 (Ludhiana); Pusa 16 Pusa 20 and Pusa 24 (IARI, New Delhi); VL Soya 2 (Almora), Shivalik (Palampur); and MACS 13 and MACS 58 (Poona).
13. Peanut improvement work is carried out by various International and National crop Research centers as follows:
14. **International Centers:** International Crop Research Institute for Semi-arid Tropics [ICRISAT], Hyderabad is involved in peanut improvement.
15. **National Centers:** In India, groundnut improvement work is carried out by following organizations.
16. Directorate of Groundnut Research, Junagarh, Gujarat.
17. State Agricultural Universities located in peanut growing regions.
18. The new varieties are released through coordinated project after multi-location testing for 3-5 years.

19. In India, more than 60 improved varieties of peanut have been released for cultivation in different. states. List of important peanut varieties released in Spanish, Valencia and Virginia groups in India is presented as follows:
20. **Spanish Type**: MH 1, SG 84,GG 2, J 11, Jyoti, TG 17, Javan, Kisan, ICGS 11, Co 1, Co 2, TMV 5, TMV 7, TMV 9, TMV 12, *etc.*
21. **Valencia Type**: MH 2, Kopergaon 3, TMV 3, *etc.*
22. **Virginia Type**: T 28, T 64, M 145, M197, TMV 6, TMV 8, TMV 10, Kausal, Kadri 2 and Kadri 3 are semi-spreading types. Spreading types include M 13, M37, M335, GG 11, TMV 1, TMV 3, TMV 4, S 230, Kadri 71-1, *etc.*
23. The seeds of *brassica* contain 40 to 45 per cent oil and 38 to 41 per cent protein. In India, oil is used for human consumption and cake as livestock feed.
24. It is believed that Rapeseed has (*Brassica campestris*) has originated in Himalayan region. Middle East is considered as primary cenre of origin of Indian Mustard. China, North East India and Caucasus are considered as secondary centres of origin. *Brassica* juncea is amphidiploid between *B. nigra* and *B. campestris*, *Brassica carinata* between *B. nigra* and *B. oleracea*, and *Brassica napus* between *B. oleracea* and *B. campestris*.
25. In *brassica*, there are three basic species, *viz.* black mustard [*Brassica nigra*] (*BB*, $2n = 16$), cabbage mustard [*B. oleracea* (*CC*, $2n = 18$)] and yellow sarson [*B. campestris* (*AA*, $2n = 20$)].
26. *Brassica napus* and *Brassica juncea* are self-pollinated species with natural outcrossing up to 30 per cent. Indian yellow sarson is self- compatible.
27. *Brassica napus* and *Brassica juncea* are self-pollinated species with natural outcrossing up to 30 per cent. Indian yellow sarson is self- compatible.
28. **Self-incompatibility**: Brown sarson and Toria are self-incompatible. Brassica nigra and *B. oleracea* are also self-incompatible. Self-incompatibility promotes cross pollination in these species which occurs by wind and honey bees.
29. **Propagation**: Rapeseed and mustard are seed propagated crops.
30. **Isolation Distance**: An isolation distance of 400 meters for the production of foundation seed and 200 meters for certified seed is required under Indian conditions.

C. MULTIPLE CHOICE QUESTIONS

1. **Which of the following is oilseed crop?**

(a) Soybean
(b) Rapeseed and Mustard
(c) Ground nut
(d) All of these

2. **Which of the following is oilseed crop?**

(a) Sunflower
(b) Safflower
(c) Castor
(d) All of these

3. **Where is Indian Institute of Soybean Research located?**

(a) Hyderabad (b) Junagarh
(c) Indore (d) All of these

4. **Where is Indian Institute of Oilseeds Research located?**

(a) Hyderabad (b) Junagarh
(c) Indore (d) All of these

5. **Where is Directorate of ground nut Research located?**

(a) Hyderabad (b) Junagarh
(c) Indore (d) All of these

6. **Where is Directorate of Rapeseed and Mustard Research located?**

(a) Bharatpur (b) Junagarh
(c) Indore (d) All of these

7. **Who proposed Triangle of origin of tetraploid species of Brassica?**

(a) Kihara (b) Nagaharu U.
(c) Tanaka (d) All of these

8. **Who sequenced Brassica genome?**

(a) Kihara (b) Nagaharu U.
(c) Wang *et al.* (d) All of these

9. **In which year Brassica genome was sequenced?**

(a) 2009 (b) 2011
(c) 2015 (d) All of these

10. **Which species of Brassica was first sequenced?**

(a) *Brassica juncea* (b) *Brassica nigra*
(c) *Brassica rapa* (d) All of these

D. TRUE OR FALSE

1. Brown sarson and Toria are self-incompatible.
2. *Brassica napus* and *Brassica juncea* are self-pollinated species.
3. Rapeseed has (*Brassica campestris*) has originated in Himalayan region.
4. *Brassica juncea* is amphidiploid between *B. nigra* and *B. campestris.*
5. *Brassica carinata* is amphidiploid between *B. nigra* and *B. oleracea.*
6. *Brassica napus* is amphidiploid between *B. oleracea* and *B. campestris.*
7. Directorate of Groundnut Research is located at Junagarh.
8. Directorate of Rapeseed and Mustard Research is located at Bharatpur.

9. Indian Institute of Soybean Research is located at Indore.
10. Indian Institute of Oilseeds Research is located at Hyderabad.
11. Indian Institute of Research on oil palm is located at Pedavegi.

E. FILL IN THE BLANKS

1. Brown sarson and Toria are _______species.
2. *Brassica napus* and *Brassica juncea* are _______ species.
3. Rapeseed has (*Brassica campestris*) originated in ______ region.
4. *Brassica juncea* is amphidiploid between *B. nigra* and _______.
5. *Brassica carinata* is amphidiploid between *B. nigra* and _______.
6. *Brassica napus* is amphidiploid between *B. oleracea* and _______.
7. Directorate of Groundnut Research is located at_______.
8. Indian Institute of Soybean Research is located at _______.
9. Indian Institute of Oilseeds Research is located at ________.
10. Indian Institute of Research on oil palm is located at _______.

F. ONE WORD OR SHORT ANSWER

1. Domesticated and commercially cultivated plants or agricultural produce, such as grain, vegetables, or fruit, considered as a group are called crop plants: Wheat is a common crop.
2. A crop (other than fruits or vegetables) that is grown for agricultural purposes is called field crop. Cotton, pulses, sugarcane wheat, mustard, maize are examples of field crops"
3. The starchy grains of various grasses such as wheat, oats, or corn, which are used as food refer to cereal crops.
4. Crops that can be used for food for humans. Examples: wheat, corn, fruits, vegetables are known as food crops.
5. Weight of 100 seeds in gram.
6. Those species of the genus Gossypium which produce spinnable seed coat fibres.
7. In groundnut, the pericarp of the pod.

ANSWERS

C. MULTIPLE CHOICE QUESTIONS

Sl.No.	*Answer*	*Sl.No.*	*Answer*
1	(d) All of these	2	(d) All of these
3	(c) Indore	4	(a) Hyderabad
5	(b) Junagarh	6	(a) Bharatpur
7	(b) Nagaharu U.	8	(c) Wang *et al.*
9	(b) 2011	10	*(c) Brassica rapa*

D. TRUE OR FALSE

1 to 11 all are true.

E. FILL IN THE BLANKS

Sl.No.	*Answer*	*Sl.No.*	*Answer*
1	Self-incompatible	2	Self-pollinated
3	Himalayan	4	B. campestris
5	B. oleracea	6	B. campestris
7	Junagarh	8	Indore
9	Hyderabad	10	Pedavegi

F. ONE WORD OR SHORT ANSWER

Sl.No.	*Answer*	*Sl.No.*	*Answer*
1	Crop Plants	2	Field Crop
3	Cereal Crops	4	Food Crops
5	Test Weight	6	Cotton
7	Shell		

Chapter 18

Breeding Pulse Crops

A. DEFINITIONS

Cereal Crops: The starchy grains of various grasses such as wheat, oats, or corn, which are used as food refer to cereal crops.

Crop Plants: Domesticated and commercially cultivated plants or agricultural produce, such as grain, vegetables, or fruit, considered as a group are called crop plants: Wheat is a common crop.

Fibre Crops: Field crops which are grown for their fibers, which are traditionally used to make paper, cloth, or rope are called fibre crops. Such crops include cotton, Jute, sun-hemp, Kenef, *etc.*

Field Crop: A crop (other than fruits or vegetables) that is grown for agricultural purposes is called field crop. Cotton, pulses, sugarcane wheat, mustard, maize are examples of field crops"

Pulse Crops: Leguminous plants producing edible seeds, *e.g.* chickpeas, lentils, beans, black gram, green gram, pigeon pea *etc.* are known as pulse crops.

Pulses: Those species of the family Leguminosae whose seeds are used for human consumption.

Test Weight: Weight of 100 seeds in gram.

B. BACKGROUND INFORMATION

1. Those Leguminous species whose seeds are used for human consumption are referred to as pulses or grain legumes (Roberts, 1970).
2. Important pulse crops include chickpea, field pea, pigeon pea, black gram, green gram, lentil, horse gram, moth bean, field bean, *khesary*, cowpea, cluster bean, *etc.* Pulse crops belong to the family Fabaceae (old Leguminosae).

3. Cool Season Pulses: Such species include chick pea, field pea, lentil and *khesary* are cool season pulses of subtropical and mild temperate regions.
4. Warm season Pulses: Such species include Pigeon pea, green gram, black gram and cowpea. Rest are warm season pulses grown in tropical and subtropical climates.
5. Based on utilization, pulse crops are again of two types, *viz.* Primary Pulses and secondary pulses. Primary Pulses include chickpea, pea, lentil (cool season), pigeon pea, green gram, black gram and cowpea (warm season). These are also known as major pulse crops.
6. (ii) Secondary Pulses include horse gram, moth bean, field bean, *khesary*, and cluster bean. These are also known as minor pulses.
7. Mutation breeding has helped in developing improved varieties in various pulse crops such as chickpea (Pusa 408, Pusa 413, Pusa 417 and Kiran), Lentil (Arun), field pea (Hans), Pigeon pea (Co 5, Co 3, TAT 5 and Trombay), black gram (Co 4), green gram (Co 4, ML 26-1-3, Pant Mung 2 and TAP 7) and cowpea (V 16, V 3, V 38 and V 240). In Pigeon pea, heterosis breeding is also used. Biotechnology will supplement to conventional methods in pulse crops.
8. **(a) ICRISAT**: International Crop Research Institute for Semiarid Tropics, Hyderabad, India is actively engaged in improvement of pigeon pea and chickpea (both desi and Kabuli).
9. **ICARDA:** International Centre for Agricultural Research in Dry-land Areas (ICARDA) works on improvement of lentil and Kabuli chickpea.
10. **AVRDC:** Asian Vegetable Research and Development Centre, Taiwan deals with breeding of mung bean and maintains global gene pool of this pulse.
11. **CIAT**: International Centre for Tropical Agriculture (CIAT), Colombia deals with breeding of French bean.
12. **IITA**: International Institute of Tropical Agriculture (IITA), Nigeria deals with improvement of Cowpea. \
13. Indian Institute of Pulses Research Kalyanpur, Kanpur: The new varieties are released through coordinated project after multi-location testing for 3-5 years. A list of important varieties released in various pulse crops in India is given below:
14. **Pigeon pea:** T 21, T 7, T 17, Prabhat, UPAS 120, Manak, Vishaka, ICPL 87, ICPL 151, LRC 30, LRC 36, Pusa 84, Pusa 85, *etc.*
15. **Chick pea:** T 3, T 4, T 5, Pant 114, K 850, H 208, H 355, Pusa 209, Pusa 240, Pusa 244, Pusa 256, Pusa 413, Pusa 417, Pusa 212, Avrodhi, *etc.*
16. **Lentil:** T 8, T 36, Pant 406, Pant L 639, L 9-12, Lens 4076, Malika, Pusa 4, Pusa 6, *etc.*

17. **Pea:** Type 163, Pant P 5, Rachna, DMR 11, Pant P 5, Hans, VL 1, B 12 *etc.*
18. **Black gram:** T 9, T 27, T 65, PS 1, Pant 419, 430, Khargone 3, Co 1, Co 2, Co 3, Co 4, Co 5, *etc.*
19. **Green gram:** P 37, P 310, PS 16, Pusa 105, Pusa Baisakhi, T 44, T 51, K 851, Plant Moong 2, Plant Moong 3, *etc.*
20. **Cowpea:** Pusa 152, Pusa Sawani, V 16, V 240, Co 1, Co 2, Co 3, Co 4, K 11, K 14, T 2, JC 5, JC 10, *etc.*

C. MULTIPLE CHOICE QUESTIONS

1. Which of the following is called primary pulse crop?

(a) Chick pea (b) Pea
(c) Lentil (d) All of these

2. Which of the following is called primary pulse crop?

(a) Pigeon pea (b) Green gram
(c) Black gram (d) All of these

3. Which of the following is a cool season pulse crop?

(a) Chick pea (b) Pea
(c) Lentil (d) All of these

4. Which of the following is a warm season pulse crop?

(a) Pigeon pea (b) Green gram
(c) Black gram (d) All of these

5. Which of the following is a secondary pulse crop?

(a) Horse gram (b) Moth bean
(c) Lathyrus (d) All of these

6. Which of the following is a secondary pulse crop?

(a) Field bean (b) Moth bean
(c) Cluster bean (d) All of these

7. Where is ICRISAT located?

(a) Kanpur (b) Hyderabad
(c) Ludhiana (d) None of these

8. Where is IIPR located?

(a) Kanpur (b) Hyderabad
(c) Ludhiana (d) None of these

9. Where is IITA located?

(a) Colombia
(b) Syria
(c) Nigeria
(d) None of these

10. Where is ICARDA located?

(a) Colombia
(b) Syria
(c) Nigeria
(d) None of these

D. TRUE OR FALSE

1. Primary pulses are also known as major pulses.
2. Secondary pulses are also known as minor pulses.
3. Indian Institute of Pulse Research is located at Kanpur.
4. International Centre for Tropical Agriculture (CIAT) is located in Colombia.
5. International Institute of Tropical Agriculture (IITA) is located in Nigeria.
6. International Crop Research Institute for Semiarid Tropics (ICRISAT) is located at Hyderabad.
7. International Centre for Agricultural Research in Dry land Areas (ICARDA) is located in Syria.
8. Asian Vegetable Research and Development Centre is located in Taiwan.
9. The first Pigeon pea hybrid is ICPH 8.
10. The first Pigeon pea hybrid ICPH 8 was released from ICRISAT.

E. FILL IN THE BLANKS

1. Primary pulses are also known as _____ pulses.
2. Secondary pulses are also known as ______ pulses.
3. Indian Institute of Pulse Research is located at ______.
4. International Centre for Tropical Agriculture (CIAT) is located in ______.
5. International Institute of Tropical Agriculture (IITA) is located in _____
6. International Crop Research Institute for Semiarid Tropics (ICRISAT) is located at ______.
7. International Centre for Agricultural Research in Dry land Areas (ICARDA) is located in ______.
8. Asian Vegetable Research and Development Centre is located in _____.
9. The first Pigeon pea hybrid is ________.
10. The first Pigeon pea hybrid ICPH 8 was released from ________.

F. ONE WORD OR SHORT ANSWER

1. Domesticated and commercially cultivated plants or agricultural produce, such as grain, vegetables, or fruit, considered as a group are called crop plants: Wheat is a common crop.
2. A crop (other than fruits or vegetables) that is grown for agricultural purposes is called field crop. Cotton, pulses, sugarcane wheat, mustard, maize are examples of field crops"
3. The starchy grains of various grasses such as wheat, oats, or corn, which are used as food refer to cereal crops.
4. Leguminous plants producing edible seeds, *e.g.* chickpeas, lentils, beans, black gram, green gram, pigeon pea *etc.* are known as pulse crops.
5. Field crops which are grown for their fibers, which are traditionally used to make paper, cloth, or rope are called fibre crops. Such crops include cotton, Jute, sunnhamp, Kenef, *etc.*
6. Weight of 100 seeds in gram.
7. Those species of the family Leguminosae whose seeds are used for human consumption.

ANSWERS

C. MULTIPLE CHOICE QUESTIONS

Sl.No.	Answer	Sl.No.	Answer
1	(d) All of these	2	(d) All of these
3	(d) All of these	4	(d) All of these
5	(d) All of these	6	(d) All of these
7	(b) Hyderabad	8	(a) Kanpur
9	(c) Nigeria	10	(b) Syria

D. TRUE OR FALSE

1 to 10 all are true.

E. FILL IN THE BLANKS

Sl.No.	Answer	Sl.No.	Answer
1	Major	2	Minor
3	Kanpur	4	Colombia
5	Nigeria	6	Hyderabad
7	Syria	8	Taiwan
9	ICPH 8	10	ICRISAT

F. ONE WORD OR SHORT ANSWER

Sl.No.	Answer	Sl.No.	Answer
1	Crop Plants	2	Field Crop
3	Cereal Crops	4	Pulse Crops
5	Fibre Crops	6	Test Weight
7	Pulses		

Section IV

Miscellaneous Topics

Chapter 19

Ideotype Breeding

A. DEFINITIONS

Biological yield: Total dry matter production per plant.

Harvest Index: The ratio of economic yield to the biological yield or the ratio of economic produce to the total biomass.

Ideotype: A plant model which is expected to yield greater quantity of grains, fibre, oil or other useful product when developed as a cultivar.

Ideotype breeding: A method of crop improvement which is used to enhance genetic yield potential through genetic manipulation of individual plant character.

B. BACKGROUND INFORMATION

1. In broad sense an ideotype is a biological model which is expected to perform or behave in a predictable manner within a defined environment.
2. More specifically, crop ideotype is a plant model which is expected to yield greater quantity of grains, fibre, oil or other useful product when developed as a cultivar.
3. The term ideotype was first proposed by Donald in 1968 working on wheat crop.
4. Ideotype is a moving goal which changes according to climatic situations, type of cultivation, national policy, market requirement *etc.*
5. In ideotype breeding, emphasis is given on individual morphological and physiological trait which enhances the yield. The value of each character is specified before initiating the breeding work.
6. Ideotype breeding makes use of genetically controlled physiological variation in increasing crop yields, besides various agronomic traits.

7. There are several factors which affect development of ideal plant type. Ideotype differs based on crop species, cultivation practices, socio-economic conditions of farmers and economic use of the plant parts.
8. Ideotype breeding consists of four important steps, *viz.* (1) development of conceptual theoretical model, (2) selection of base material, (3) incorporation of desirable characters into single genotype, and (4) selection of ideal or model plant type.
9. Semi-dwarf varieties of wheat and rice are highly responsive to water use and nitrogen application and have wide adaptation. These qualities have made them popular throughout the world.
10. The concept of plant type was introduced in rice breeding by Jennings in 1964, though the term ideotype was coined by Donald in 1968.
11. In 1975, Mock and Pearce proposed ideal plant type of maize.
12. Rasmusson (1987) proposed ideal plant type of six rowed barley.
13. Singh *et al.* (1974) proposed an ideal plant type of upland cotton (*Gossypium hirsutum*) and tree cotton (*G. arboreum*) for irrigated conditions of North Indian cotton growing belt.
14. Singh and Narayanan (1993) proposed an ideotype of upland and tree cottons for rain-fed conditions.
15. Ideotype breeding has significantly contributed to enhanced yields in cereals (wheat and rice) and millets (*Sorghum* and pearl millet) through the use of dwarfing genes, resulting in green revolution.
16. The Norin 10 in wheat and Dee-geo-Woo-gen in rice are the sources of dwarfing genes.

C. MULTIPLE CHOICE QUESTIONS

1. Who first developed the concept of plant ideotype?

(a) Jennings (b) Donald
(c) Rasmusson (d) None of these

2. With which crop the concept of plant ideotype was first developed?

(a) Wheat (b) Rice
(c) Barley (d) None of these

3. Who first proposed plant type of Rice?

(a) Jennings (b) Donald
(c) Rasmusson (d) None of these

4. Who first proposed plant type of six rowed barley?

(a) Jennings (b) Donald
(c) Rasmusson (d) None of these

5. Who first proposed ideal plant type of Maize?

(a) Jennings (b) Mock and Pearce

(c) Rasmusson (d) None of these

6. Who first proposed ideal plant type of Cotton?

(a) Jennings (b) Singh *et al.*

(c) Rasmusson (d) None of these

7. Who coined the term plant Ideotype?

(a) Jennings (b) Mock and Pearce

(c) Donald (d) None of these

8. Who coined the term harvest index?

(a) Jennings (b) Donald

(c) Rasmusson (d) None of these

9. What is the source of dwarfing gene in wheat?

(a) Norin 10 (b) Dee-Geo-Woogen

(c) Norin 5 (d) None of these

10. What is the source of dwarfing gene in Rice?

(a) Norin 10 (b) Dee-Geo-Woogen

(c) Norin 5 (d) none of these

D. TRUE OR FALSE

1. The concept of plant ideotype was developed by Donald.
2. The term plant ideotype was coined by Donald.
3. The term harvest index was coined by Donald.
4. The plant type of Rice was first proposed by Jennings.
5. The plant type of Maize was first proposed by Mock and Pearce.
6. The plant type of six rowed barley was first proposed by Rasmusson.
7. The concept of plant ideotype was developed with wheat crop.
8. The source of dwarfing gene in wheat is Norin 10.
9. The source of dwarfing gene in Rice is Dee-Geo-Woo-Gen.
10. IR 8 is the semi-dwarf variety of Rice.

E. FILL IN THE BLANKS

1. The concept of plant ideotype was developed by _______.
2. The term plant ideotype was coined by _______.
3. The term harvest index was coined by ______.

4. The plant type of Rice was first proposed by ______.
5. The plant type of Maize was first proposed by ______.
6. The plant type of six rowed barley was first proposed by ______.
7. The concept of plant ideotype was developed with ______ crop.
8. The source of dwarfing gene in wheat is ______.
9. The source of dwarfing gene in Rice is ______.
10. IR 8 is the semi-dwarf variety of ______.

F. ONE WORD OR SHORT ANSWER

1. A plant model which is expected to yield greater quantity of grains, fibre, oil or other useful product when developed as a cultivar.
2. A method of crop improvement which is used to enhance genetic yield potential through genetic manipulation of individual plant character.
3. The ratio of economic yield to the biological yield or the ratio of economic produce to the total biomass.
4. Total dry matter production per plant.

ANSWERS

C. MULTIPLE CHOICE QUESTIONS

Sl.No.	Answer	Sl.No.	Answer
1	(b) Donald	2	(a) Wheat
3	(a) Jennings	4	(c) Rasmusson
5	(b) Mock and Pearce	6	(b) Singh *et al.*
7	(c) Donald	8	(b) Donald
9	(a) Norin 10	10	(b) Dee-Geo-Woogen

D. TRUE OR FALSE

1 to 10 all are True.

E. FILL IN THE BLANKS

Sl.No.	Answer	Sl.No.	Answer
1	Donald	2	Donald
3	Donald	4	Jennings
5	Mock and Pearce	6	Rasmusson
7	Wheat	8	Norin 10
9	Dee-Geo-Woogen	10	Rice

F. ONE WORD OR SHORT ANSWER

Sl.No.	Answer	Sl.No.	Answer
1	Ideotype	2	Ideotype breeding
3	Harvest Index	4	Biological yield

Chapter 20

Breeding for Climate Change

A. DEFINITIONS

Climate: The long term weather or atmospheric conditions of a place.

Climate change: Any long-term significant change in the "average weather" of a region or the earth as a whole.

Weather: The day to day or short term atmospheric conditions [temperature, precipitation and wind patterns] of a place.

B. BACKGROUND INFORMATION

1. Climate change is any long-term significant change in the "average weather" of a region or the earth as a whole.
2. Average weather may include average temperature, precipitation and wind patterns. It involves changes in the variability or average state of the atmosphere over durations ranging from decades to millions of years. These changes can be caused by dynamic processes on Earth, external forces including variations in sunlight intensity, and more recently by human activities.
3. Climate change is the result of several factors such as (i) glaciation, (ii) ocean variability, (iii) CO_2 concentration, (iv) plate-tectonics, (v) solar variations, (vi) orbital variations, (vii) volcanism, (viii) fossil fuel, (ix) cement manufacture, (x) land use (irrigation and deforestation), (xi) livestock *etc.*
4. The climate change has two types of effects, *viz.* direct effects and indirect effects. The direct effects of climate change include effect on (i) temperature, (ii) rainfall, (iii) sunlight, (iv) CO_2 concentration, *etc.*

5. Gradual change in the climate over a long period will lead to several new problems such as appearance of new insects, new diseases, new weeds, high temperature, drought conditions, change in rainfall *etc.*
6. In crop plants, there are five important sources which can be used for developing suitable cultivars for cultivation in the region where climate change is taking place. The sources include (i) germplasm collection, (ii) cultivated varieties, (iii) wild species and relatives of crop plants, (iv) Induced mutations, and (v) trans-genes.
7. Traditional breeding approaches will continue to develop cultivars suitable for climate change. These approaches include introduction, selection, hybridization and mutations. In developing suitable crop cultivars for climate change due importance should be given to: (i) development of MAR lines, (ii) genetic enhancement, (iii) transgenic technology, (iv) durable resistance, and (v) crop adaptation.

C. MULTIPLE CHOICE QUESTIONS

1. Climate change in a region affects:

(a) Rainfall pattern (b) Temperature pattern
(c) Sun shine pattern (d) All of these

2. Climate change in a region affects:

(a) Disease scenario (b) Insect scenario
(c) Weed scenario (d) All of these

3. Climate change in a region affects:

(a) Biodiversity pattern (b) Insect scenario
(c) Weed scenario (d) All of these

4. Which of the following factors leads to Climate change?

(a) Glaciation (b) Ocean variability
(c) Co2 concentration (d) All of these

5. Which of the following factors leads to Climate change?

(a) Plate tectonics (b) Solar variations
(c) Orbital variations (d) All of these

6. Which of the following factors leads to Climate change?

(a) Volcanism (b) Fossil fuel
(c) Cement manufacture (d) All of these

7. What are the sources to breed for Climate change?

(a) Cultivated Varieties (b) Germplasm
(c) Wild species (d) All of these

8. What are the sources to breed for Climate change?

(a) Land Races (b) Induced mutations

(c) Transgenes (d) All of these

9. What are the sources to breed for Climate change?

(a) Land Races (b) Induced mutations

(c) Transgenes (d) All of these

10. Which of the following breeding approach will cope-up with Climate change?

(a) Development of multiple adversity resistant lines.

(b) Genetic enhancement

(c) Transgenic technology

(d) All of these

D. TRUE OR FALSE

1. Long-term significant change in the "average weather" of a region is called climate change.
2. Climate change is a long term gradual process.
3. Global warming affects the Climate change.
4. Pest scenario of a region is affected due to Climate change.
5. Disease scenario of a region is affected due to Climate change.
6. Biodiversity pattern of a region is affected due to Climate change.
7. Rainfall pattern of a region is affected due to Climate change.
8. Temperature pattern of a region is affected due to Climate change.
9. Cropping pattern of a region is affected due to Climate change.
10. Human activity is contributing to the current rapid changes in the world's climate.

E. FILL IN THE BLANKS

1. Long-term significant change in the "average weather" of a region is called ________.
2. Climate change is a long term ________ process.
3. Global warming affects the ________.
4. Pest scenario of a region is affected due to ________.
5. Disease scenario of a region is affected due to ________.
6. Biodiversity pattern of a region is affected due to ________.
7. Rainfall pattern of a region is affected due to ________.

8. Temperature pattern of a region is affected due to ________.
9. Cropping pattern of a region is affected due to __________.
10. Human activity is contributing to the current rapid _________ in the world's climate.

F. ONE WORD OR SHORT ANSWER

1. The long term weather or atmospheric conditions of a place.
2. Any long-term significant change in the "average weather" of a region or the earth as a whole.
3. The day to day or short term atmospheric conditions [temperature, precipitation and wind patterns] of a place.

ANSWERS

C. MULTIPLE CHOICE QUESTIONS

Sl.No.	Answer	Sl.No.	Answer
1	(d) All of these	2	(d) All of these
3	(d) All of these	4	(d) All of these
5	(d) All of these	6	(d) All of these
7	(d) All of these	8	(d) All of these
9	(d) All of these	10	(d) All of these

D. TRUE OR FALSE

1 to 10 all are true.

E. FILL IN THE BLANKS

Sl.No.	Answer	Sl.No.	Answer
1	Climate change	2	Gradual
3	Climate change	4	Climate change
5	Climate change	6	Climate change
7	Climate change	8	Climate change
9	Climate change	10	Changes

F. ONE WORD OR SHORT ANSWER

Sl.No.	Answer	Sl.No.	Answer
1	Climate	2	Climate change
3	Weather		

Appendices

APPENDIX 1. LIST OF AGRICULTURAL UNIVERSITIES IN INDIA

Sl.No.	Name of University	Location	State
	I. AGRICULTURAL UNIVERSITIES		
A	**STATE UNIVERSITIES**		
1	Acharya NG Ranga Agricultural University	Hyderabad	Telangana
2	Assam Agricultural University	Jorhat	Assam
3	Bihar Agricultural University	Sabour	Bihar
4	Birsa Agricultural University	Ranchi	Jharkhand
5	Indira Gandhi Krishi Vishwavidyalaya	Raipur	Chhatishgarh
6	Anand Agricultural University	Anand	Gujarat
7	Junagadh Agricultural University	Junagarh	Gujarat
8	Navsari Agricultural University	Navsari	Gujarat
9	Sardarkrushinagar-Dantiwada Agricultural University	Bansakantha	Gujarat
10	Chaudhary Charan Singh Haryana Agricultural University	Hisar	Haryana
11	CSK Himachal Pradesh Krishi Vishvavidyalaya,	Palampur	Himachal Pradesh
12	Sher-E-Kashmir Univ of Agricultural Sciences and Technology	Jammu	Jammu and Kashmir
13	Sher-E-Kashmir Univ of Agricultural Sciences and Technology of Kashmir	Srinagar	Jammu and Kashmir
14	University of Agricultural Sciences	Bangalore	Karnataka
15	University of Agricultural Sciences	Dharwad	Karnataka
16	University of Agricultural Sciences	Raichur	Karnataka
17	Kerala Agricultural University	Vallanikara	Kerela
18	Jawaharlal Nehru Krishi Viswavidyalaya	Jabalpur	Madhya Pradesh
19	Rajmata Vijayaraje Scindia Krishi Viswavidyalaya	Gwalior	Madhya Pradesh
20	Dr Balasaheb Sawant Konkan Krishi Vidyapeeth	Ratnagiri	Maharashtra
21	Dr Panjabrao Deshmukh Krishi Vidyapeeth	Akola	Maharashtra
22	Mahatma Phule Krishi Vidyapeeth	Rahuri	Maharashtra
23	Marathwada Agricultural University	Parbhani	Maharashtra

Sl.No.	*Name of University*	*Location*	*State*
24	Odisha Univ of Agriculture and Technology	Bhubaneswar	Odisha
25	Punjab Agricultural University	Ludhiana	Punjab
26	Maharana Pratap Univ of Agriculture and Technology	Udaipur	Rajasthan
27	Rajasthan Agricultural University	Bikaner	Rajasthan
28	Agriculture University	Kota	Rajasthan
29	Sri Karan Narendra Agriculture University	Jobner	Rajasthan
30	Agriculture University	Jodhpur	Rajasthan
31	Tamil Nadu Agricultural University	Coimbatore	Tamil Nadu
32	Chandra Shekar Azad University of Agriculture and Technology	Kanpur	Uttar Pradesh
33	Narendra Deva University of Agriculture and Technology	Faizabad	Uttar Pradesh
34	Sardar Ballabh Bhai Patel Univ of Agriculture and Technology	Meerut	Uttar Pradesh
35	Manyavar Kashiram University of Agriculture and Technology	Banda	Uttar Pradesh
36	Govind Ballabh Pant University of Agriculture and Technology	Pantnagar	Uttrakhand
37	Bidhan Chandra Krishi Viswavidyalaya	Kalyani	West Bengal
38	Uttar Banga Krishi Viswavidyalaya	Cooch Bihar	West Bengal
B	**CENTRAL UNIVERSITIES**		
1	Central Agricultural University	Imphal	Manipur
2	Rani Laxmibai Central Agricultural University	Jhansi	Uttar Pradesh
3	Rajendra Central Agricultural University	Pusa	Bihar
C	**DEEMED UNIVERSITIES**		
1	Indian Agricultural Research Institute	Pusa	New Delhi
2	National Dairy Research Institute	Karnal	Haryana
3	Indian Veterinary Research Institute	Bareilly	Uttar Pradesh
4	Central Institute for Fisheries Education	Mumbai	Maharashtra
5	Allahabad Agricultural Institute	Allahabad	Uttar Pradesh
II. ANIMAL SCIENCE UNIVERSITIES			
S. N.	Name of University	Location	State
1	Sri Venkateswara Veterinary University	Tirupati	Andhra Pradesh

Sl.No.	*Name of University*	*Location*	*State*
2	Karnataka Veterinary, Animal and Fisheries Sciences University	Bidar	Karnataka
3	Maharashtra Animal Science and Fishery University	Nagpur	Maharashtra
4	Guru Angad Dev Veterinary and Animal Science University	Ludhiana	Punjab
5	Tamil Nadu Veterinary and Animal Science University	Chennai	Tamil Nadu
6	Tamil Nadu Fisheries University	Nagapattinam	Tamil Nadu
7	UP Pandit Deen Dayal Upadhaya Pashu Chikitsa Vigyan Vishwa Vidhyalaya evam Go Anusandhan Sansthan, Mathura	Mathura	Uttar Pradesh
8	West Bengal University of Animal and Fishery Sciences	Kolkata	West Bengal
9	Rajasthan University of Veterinary and Animal Sciences	Bikaner	Rajasthan
10	Kerala University of Fisheries and Ocean Studies	Kochi	Kerala
11	Kerala Veterinary and Animal Sciences University	Wayanand	Kerala
III. HORTICULTURAL UNIVERSITIES			
1	Andhra Pradesh Horticultural University	Tadepalli-gudem	Andhra Pradesh
2	University of Horticultural Sciences	Bagalkot	Karnataka
3	Dr Yashwant Singh Parmar Univ of Horticulture and Forestry	Solan	Himachal Pradesh
4	Dr YSR Horticultural University	Veketara-manna-gudem	Andhra Pradesh
5	Uttarakhand University of Horticulture and Forestry	Bharsar	Uttarakhand

APPENDIX 2. LIST OF ICAR RESEARCH INSTITUTES IN INDIA

Sl.No.	Name of Research Institute	Location	State
A	**CROP SCIENCES**		
1	ICAR-Central Institute for Cotton Research	Nagpur	Maharashtra
2	ICAR- Central Institute for Jute and Allied Fibres	Kolkata	West Bengal
3	ICAR-National Rice Research Institute	Cuttack	Odisha
4	ICAR -Central Tobacco Research Institute	Rajamundry	Andhra Pradesh
5	ICAR- Directorate of Groundnut Research	Junagarh	Gujarat
6	ICAR-Indian Institute of Oilseeds Research	Hydereabad	Telangana
7	ICAR- Directorate of Rapeseed-Mustard Research	Bharatpur	Rajasthan
8	ICAR-Indian Institute of Rice Research	Hyderabad	Talangana
9	ICAR-Indian Institute of Seed Science	Mau	Uttar Pradesh
10	ICAR-Indian Institute of Millets Research	Hyderabad	Talangana
11	ICAR-Indian Institute of Soybean Research	Indore	Madhya Pradesh
12	ICAR-Indian Agriculture Research Institute	Pusa	New Delhi
13	ICAR-Indian Agriculture Research Institute	Hazaribagh	Jharkhand
14	ICAR-Indian Grassland and Fodder Research Institute	Jhansi	Uttar Pradesh
15	ICAR-Indian Institute of Maize Research	Ludhiana	Punjab
16	ICAR-Indian Institute of Agricultural Biotechnology	Ranchi	Bihar
17	ICAR-Indian Institute of Pulses Research	Kanpur	Uttar Pradesh
18	ICAR-Indian Institute of Sugarcane Research	Lucknow	Uttar Pradesh
19	ICAR-Indian Institute of Wheat and Barley Research	Karnal	Haryana
20	ICAR-National Bureau of Agricultural Insect Resources	Bangaluru	Karnataka

Sl.No.	Name of Research Institute	Location	State
21	ICAR-National Bureau of Agriculturally Important Micro-organism	Mau	Uttar Pradesh
22	ICAR-National Bureau of Plant Genetic Resources	Pusa	New Delhi
23	ICAR-National Centre for Integrated Pest Management	Pusa	New Delhi
24	ICAR- National Institute of Biotic Stress Management	Raipur	Chhattishgarh
25	ICAR-National Centre on Plant Biotechnology	Pusa	New Delhi
26	ICAR-Sugarcane Breeding Institute	Coimbatore	Tamil Nadu
27	ICAR-Vivekanand Parvatiya Krishi Anusandhan Sansthan	Almora	Uttrakhand
B	**HORTICULTURAL SCIENCES**		
28	ICAR-Central Institute for Arid Horticulture	Bikaner	Rajasthan
29	ICAR-Central Institute for Subtropical Horticulture	Lucknow	Uttar Pradesh
30	ICAR-Central Institute for Temperate Horticulture	Srinagar	Jammu and Kashmir
31	ICAR- Central Island Agricultural Research Institute	Portblair	Andman and Nicobar
32	ICAR-Central Plantation Crops Research Institute	Kasargod	Kerala
33	ICAR-Central Potato Research Institute	Shimla	Himachal Pradesh
34	ICAR-Central Tuber Crops Research Institute	Trivendrum	Kerala
35	ICAR-Directorate of Cashew Research	Puttur	Karnataka
36	ICAR-Directorate of Floriculture Research	Pusa	New Delhi
37	ICAR-Directorate of Medicinal And Aromatic Plants Research	Anand	Gujarat
38	ICAR-Directorate of Mashroom Research	Solan	Himachal Pradesh
39	ICAR-Directorate of Onion and Garlic Research	Pune	Maharashtra
40	ICAR-Indian Institute of Horticulture Research	Bangalore	Karnataka

Sl.No.	Name of Research Institute	Location	State
41	ICAR-Indian Institute of Oil Palm Research	Pedavegi	Andhra Pradesh
42	ICAR-Indian Institute of Spices Research	Calicut	Kerala
43	ICAR-Indian Institute of Vegetable Research	Varanasi	Uttar Pradesh
44	ICAR-National Research Centre for Banana	Tiruchirapalli	Kerala
45	ICAR-Central Citrus Research Institute	Nagpur	Maharashtra
46	ICAR-National Research Centre for Grapes	Pune	Maharashtra
47	ICAR-National Research Centre for Litchi	Muzaffarpur	Bihar
48	ICAR-National Research Centre for Orchids	Gangtok	Sikkim
49	ICAR-National Research Centre on Pomegranate	Solapur	Maharashtra
C	**NATURAL RESOURCE MANAGEMENT**		
50	ICAR-National Research Centre for Seed Spices	Ajmer	Rajasthan
51	ICAR-Central Agro-forestry Research Institute	Jhansi	Uttar Pradesh
52	ICAR-Central Arid Zone Research Institute	Jodhpur	Rajasthan
53	ICAR-Central Research Institute for Dry land Agriculture	Hyderabad	Andhra Pradesh
54	ICAR-Central Soil and Water Conservation	Dehradun	Uttrakhand
55	ICAR-Central Soil Salinity Research Institute	Karnal	Haryana
56	ICAR-Indian Institute of Water Management	Bhubaneswar	Odisha
57	ICAR-Directorate of Weed Research	Jabalpur	Madhya Pradesh
58	ICAR Research Complex for NEH Region	Shillong	Meghalaya
59	ICAR Research Complex for Eastern Region	Patna	Bihar
60	ICAR-Indian Institute of Soil Science	Bhopal	Madhya Pradesh

Sl.No.	Name of Research Institute	Location	State
61	ICAR-Indian Institute of Farming System Research	Meerut	Uttar Pradesh
62	ICAR-Indian Institute of Soil Science	Bhopal	Madhya Pradesh
63	ICAR-National Bureau of Soil Survey and Land Use Planning	Nagpur	Maharashtra
64	ICAR-National Institute of Abiotic Research management	Baramati	Maharashtra
65	ICAR- National Organic Farming Research	Gangtok	Sikkim
66	ICAR_ National Research Centre for Integrated Farming	Morihari	Bihar
D	**AGRICULTURAL ENGINEERING**		
67	ICAR-Central Institute for Research on Cotton Technology	Mumbai	Maharashtra
68	ICAR-Central Institute of Agricultural Engineering	Bhopal	Madhya Pradesh
69	ICAR-Central Institute Post Harvest Engineering and Technology	Ludhiana	Punjab
70	ICAR-Indian Institute of Natural Resigns and Gums	Ranchi	Jharkhand
71	ICAR-National Institute of Research on Jute and Allied Fibres Technology	Kolkata	West Bengal
E	**AGRICULTURAL EDUCATION**		
72	ICAR-Central Institute for Women in Agriculture	Bhubaneswar	Odisha
73	ICAR-Indian Agricultural Statistics Research Institute	Pusa Campus	New Delhi
74	ICAR-National Academy of Agricultural Research Management	Hyderabad	Telangana
75	ICAR-National Institute of Agricultural Economics and Policy Research	Pusa Campus	New Delhi
F	**AGRICULTURAL EXTENSION**		
76	ICAR-Directorate of Knowledge Management in Agriculture	Pusa Campus	New Delhi
G	**ANIMAL RESEARCH INSTITUTES**		
77	ICAR-Central Avian Research Institute	Bareilly	Uttar Pradesh
78	ICAR-Central Institute for Research on Buffaloes	Hisar	Haryana

Sl.No.	Name of Research Institute	Location	State
79	ICAR-Central Institute for Research on Cattle	Meerut	Uttar Pradesh
80	ICAR-Central Institute for Research on Goats	Mathura	Uttar Pradesh
81	ICAR-Central Sheep and Wool Research Institute	Avikanagar	Rajasthan
82	ICAR-Directorate of Poultry Research	Hyderabad	Telangana
83	ICAR-Indian Veterinary Research Institute	Bareilly	Uttar Pradesh
84	ICAR-National Bureau of Animal Genetic Resources	Karnal	Haryana
85	ICAR-National Dairy Research Institute	Karnal	Haryana
86	ICAR-National Institute of Animal Nutrition and Physiology	Karnal	Haryana
87	ICAR-National Institute of High Security Animal Diseases	Bhopal	Madhya Pradesh
88	ICAR-National Institute of Veterinary Epidemiology and Disease Informatics	Bangalore	Karnataka
89	ICAR-National Research Centre for Equines	Hisar	Haryana
90	ICAR-National Research Centre for Camel	Bikaner	Rajasthan
91	ICAR-National Research Centre on Meat	Hyderabad	Telangana
92	ICAR-National Research Centre on Mithun	Jharnapani	Nagaland
93	ICAR-National Research Centre on Pig	Guwahati	Assam
94	ICAR-National Research Centre on yak	Dirang	Arunachal Pradesh
95	ICAR-Project Directorate on Foot and Mouth Disease	Nainital	Uttrakhand
H	**FISHERIES INSTITUTES**		
96	ICAR-Central Island Fisheries Research Institute	Barrackpur	West Bengal
97	ICAR-Central Institute of Brackish water Aquaculture	Chennai	Tamil Nadu
98	ICAR-Central Institute Fisheries Education	Mumbai	Maharashtra
99	ICAR-Central Institute Fisheries Technology	Cochin	Kerala

Sl.No.	Name of Research Institute	Location	State
100	ICAR-Central Institute of Fresh Water Aquaculture	Bhubaneswar	Odisha
101	ICAR-Central Marine Fisheries Research Institute	Kochi	Kerala
102	ICAR-Directorate of Cold Water Fisheries Research	Bhimtal	Uttarakhand
103	ICAR-National Bureau of Fish Genetic Resources	Lucknow	Uttar Pradesh

APPENDIX 3: LIST OF INTERNATIONAL AGRICULTURAL RESEARCH INSTITUTES/CENTRES

Sl. No.	*Name of Research Institute/Centre*	*Founded in*	*Location*	*Country*
1.	International Rice Research Institute [IRRI]	1960	Manila	Philippines
2.	International Wheat and Maize Improvement Centre [CIMMYT]	1963	Mexico City	Mexico
3.	International Centre for Tropical Agriculture [CIAT]	1967	Cali	Colombia
4.	International Institute for tropical Agriculture [IITA]	1967	Ibadan	Nigeria
5.	International Potato Centre [CIP]	1971	Lima	Peru
6.	International Crop Research Institute for Semiarid Tropics [ICRISAT]	1972	Hyderabad	India
7.	Internationa Centre for Agriculture Research in Drylans Areas [ICARDA]	1977	Aleppo	Syria
8.	West Africa Rice Development Association [WARDA]	1970	Bouake	Cote divoire
9.	International Plant Genetic Resources Institute [IPGRI]	1974	Rome	Italy
10.	Centre for International Forestry Research [CIFOR]	1992	Jakarta	Indonesia
11.	International Centre for Research in Agroforestry [ICRAF]	1991	Nairobi	Kenya
12.	International Centre for Living Aquati Resources Management [ICLARM]	1977	Makati City	Philippines
13.	International Food Policy Research Institute [IFPRI]	1975	Washington DC	USA
14.	International Irrigation Management Institute [IIMI]	1984	Colombo	Sri Lanka
15.	International Livestock Research Institute [ILRI]	1995	Nairobi	Kenya
16.	International service for National Agricultural research [ISNAR]	1979	Hague	Netherlands

Sl. No.	*Name of Research Institute/Centre*	*Founded in*	*Location*	*Country*
17.	Asian Vegetable Research and Development Centre [AVRDC]	1971	Taiwan City	Taiwan [China]
18.	International Centre for Genetic Engineering and Biotechnology [ICGEB]	1987	Trieste	Italy

Glossary

CHAPTER 1: INTRODUCTION AND OBJECTIVES

Abiotic Stress: Adverse conditions imposed for growth and production of crop plants by environmental factors such as deficiency or excess of nutrition, moisture, temperature and light; presence of harmful gases or toxicants; and abnormal soil conditions, such as salinity, alkalinity and acidity.

Adaptability: Stable performance of a variety over a wide range of environmental conditions.

Biotic Stress: Adverse conditions imposed for growth and production of crop plants by biotic factors such as insects, diseases and parasitic weeds.

Breeding Techniques: Various breeding procedures which are used for genetic improvement of crop plants in relation to their economic use.

Easy Care Property: In cotton. fabrics that are easily washed and require little pressing.

Genetic Resistance: Ability of some genotypes to give higher yield of good quality than other varieties at the same initial level of disease or insect infestation under similar environmental conditions.

Germplasm: The whole library of alleles in a crop species or sum total of genes in a species.

Harvest Index: The ratio of economic yield to biological yield. Biological yield is the total dry matter production per plant.

Plant Breeding: A science, an art and a technology which deals with genetic improvement of crop plants in relation to their economic use for mankind, also called as crop improvement.

Quality: Suitability or fitness of an economic plant product in relation to its end use.

Seed Production Technology: A branch of plant breeding which deals with principles and methods of improved seed production.

CHAPTER 2: CENTRES OF ORIGIN AND DISTRIBUTION OF CROP SPECIES

Biodiversity: Total variability present within and among species of all living organisms and their habitats.

Centres of Diversity: A place, region or area where maximum variability of crop plants is observed; also called centres of origin.

Extinction: Permanent loss of a crop species due to various reasons.

Field Gene Banks: Those areas of land in which germplasm of recalcitrant crop species is maintained in the form of plants; also called plant gene banks.

Gene Banks: Various organizations where genetic diversity is maintained in living state; also called as germplasm banks.

Gene Sanctuaries: Protected areas of great genetic diversity under natural conditions *i.e. in situ* conservation.

Genetic Diversity: Total amount of genetic variation present in a population or species or variety of genes and genotypes found in a particular crop species.

Genetic Erosion: Loss of genetic diversity between and within population of a species over a period of time.

Law of Parallel Variation: This states that a particular variation observed in a crop species is also expected to be available in its another related species. Also called law of homologous series of variation.

Micro Centres: Small areas within the centre of diversity that exhibits tremendous genetic diversity of crop plants.

Plant genetic Resources: Genetic material of plants which is of value as a resource for present and future generations of people.

Primary Centre of Diversity: The original homes of crop plants which are generally uncultivated areas such as mountains, hills, river valleys, forests *etc.*

Secondary Centres of Diversity: The cultivated areas with vast genetic diversity of a species.

Seed gene banks: Conservation of germplasm in the form of seeds in cold storage.

CHAPTER 3: PLANT GENETIC RESOURCES: CONSERVATION AND UTILIZATION

Active Collections: Collections which are actively utilized in the breeding programmes and are used for medium term storage (8-10 years). Seeds are stored at 0°C.

Base Collections: Total accessions available in a crop. These are used for long storage (upto 100 years). Seeds are stored at -18 or -20°C; also called principal collections.

Characterization: Recording of highly heritable phenotypic characters.

Conservation: The protection of genetic diversity from genetic erosion either under natural conditions or by storing in gene banks.

Core Collections: A set of accession derived from base collections to represent the genetic spectrum in the whole collection.

Documentation: The process of compilation, analysis, classification, storage and distribution of information.

***Ex situ* Conservation:** Preservation of germplasm in the gene banks.

Exotic Collection: The germplasm which is collected or received from other countries.

Gene Pool: The whole library of different alleles of a species or sum total of genes in a species. Also called germplasm, genetic stock and genetic resources.

Genetic Erosion: Gradual reduction in genetic variability due to elimination of various geneotypes. It results due to use of modern cultivars, modernization of agriculture and various development activities.

***In situ* Conservation:** Conservation of germplasm under natural conditions.

Indigeneous Collection: The germplasm which is collected within the country.

Land Races. The primitive cultivars which were selected and cultivated by farmers for many generations. Land races have more genetic diversity, wider adaptability and high degree of resistance to biotic and abiotic stresses.

Modern Cultivars: The currently cultivated high yielding varieties.

Obsolete Cultivars: The improved cultivars of the recent past.

Orthodox Seeds: Seeds which can be dried to low moisture content and stored at low temperature without losing their viability.

Plant Exploration: Trips arranged for collection of germplasm from different areas.

Primary Gene Pool: The gene pool in which intermating is easy and leads to production of fertile hybrids. It includes genotypes of same species. It is designated as GP1.

Quarantine: The prophylactic measure that is used to prevent the entry of new diseases, insects and weeds from other countries.

Recalcitrant Seeds: Seeds which show very drastic loss in viability with decrease in moisture content below 12 to 13 per cent. Such species include coconut, mango, tea, coffee, rubber, jack fruit, oil palm *etc.* Such seeds cannot be conserved in seed banks.

Secondary Gene Pool: The genetic material that leads to partial fertility on crossing with primary gene pool. It includes genotypes of related species and is designated at GP2.

Tertiary Gene Pool: The genetic material which leads to production of sterile hybrids on crossing with primary gene pool. It is designated as GP3.

Working Collections: Collections which are frequently utilized by breeders in their crop improvement programs. These are stored for short term (3-5 years) at 5-10°C.

CHAPTER 4: GENETICS OF QUALITATIVE AND QUANTITATIVE CHARACTERS

Additive variance: Average effect of genes on all segregating loci. Such genes show lack of dominance, *i.e.*, intermediate expression.

Additive x Additive Epistasis: Interaction between two or more loci each exhibiting lack of dominance individually. It is denoted as *AA* and is fixable.

Additive x Dominance Epistasis: Interaction between two or more loci, one exhibiting lack of dominance and the other dominance. It is denoted as *AD* and is non-fixable.

Character: Any property of an individual showing heritable variation. It includes morphological, physiological, biochemical and behavioural properties.

Dominance Variance: Deviation from the mean value due to intra-allelic interaction.

Dominance x Dominance Epistasis: Interaction of two or more loci, each exhibiting dominance individually. It is represented as *DD* and is nonfixable.

Environmental Variation: The non-heritable variation which is entirely due to environmental effects and varies under different environmental conditions.

Epistatic Variance: Deviation from the mean value as a consequence of non-allelic interaction, *i.e.*, interaction between two or more genes. It is of three types *viz.*, additive x additive, additive x dominance and dominance x dominance.

Genetic Variance: Heritable portion of total or phenotypic variance. It is of three types, *viz.* additive, dominance and epistatic variances.

Genotypic Variation: The inherent or genetic variation which remains unaltered by environmental changes.

Oligogenic Trait: Characters which are controlled by one or few genes each having detectable individual effect and exhibit discontinuous variation.

Phenotypic Variation: The total variability which is observable.

Polygenic Traits: Characters which are governed by several genes each having small individual effect and exhibit discontinuous variation.

CHAPTER 5: GENETIC BASIS OF BREEDING METHODS

Allogamy: Development of seed by cross pollination.

Autogamy: Development of seed by self pollination.

Heterogeneous Population: A population that is composed of genetically dissimilar plants such as land races, mass selected populations, composites, synthetics and multilines.

Heterozygous Population: A Population that segregates on selfing such as *F*1 hybrids, composites and synthetics. Heterozygous individuals have unlike alleles at the corresponding loci.

Homogeneous Population: A population of genetically similar plants such as pure line, *F*1 between two pure lines and progeny of a clone.

Homozygosity: The proportion of homozygous individuals in a segregating population. Homozygosity = $[(2m-1)/2m]n$, where m = number of generations of self pollination and n = number of gene pairs segregating.

Homozygous Population: A true breeding population such as pure lines and mass selected populations in self-pollinated species. Homozygous individuals have identical alleles at the corresponding loci.

Plant Breeding Methods: Various procedures (Selection, hybridization, mutation, *etc.*) that are used for genetic improvement of crop plants.

CHAPTER 6: BREEDING SELF POLLINATED CROPS [INTRODUCTION AND PURE LINE SELECTION]

Acclimatization: Adaptation of an introduced variety to the new environment.

Artificial Selection: The selection made by human. It fevours those characteristics of plants that are related to yield and quality.

Characterization: Recording observations on highly heritable traits such as colour, shape, hairiness *etc.*

Cyclic Selection: Selection in one direction for one generation or season and in opposite direction in next generation or season.

Direct Introduction: Introductions which are immediately adapted to the changed environment.

Directional Selection: Selection in favour of an extreme phenotype such as earliness or lateness, dwarfness or tallness.

Disruptive Selection: Selection in favour of two extreme types, *viz.* earliness and lateness or tallness and dwarfness.

Domestication. The process of bringing wild and weedy species under human management.

Exotic Variety: A foreign variety which is directly recommended for commercial cultivation.

Indirect Introduction: Introductions which take some years for adaptation in changed environment.

Natural Selection: Selection that operates in nature without human interference. It favours those plant characters that are essential for survival (adaptation) of a species.

Plant Introduction: Transposition of crop plants from the place of their cultivation to such areas where they were never grown earlier.

Primary introduction: Introduction that can be used for commercial cultivation without any change in the original genotype.

Pure line Selection: Development of new variety through identification and isolation of a single best plant progeny.

Pure Line: Progeny of a self-pollinated homozygous plant obtained by selfing.

Secondary Introduction: Introduction that can be used as a variety after selection from original genotype or used for transfer of some desirable gene to the cultivated variety.

Selection: The process that favours survival and further propagation of some plants having more desirable characters than others. It is of two types, *viz.*, natural and artificial.

Stabilizing Selection: Selection in favour of intermediate types; also called centripetal selection.

CHAPTER 7: BREEDING SELF POLLINATED CROPS [PEDIGREE, BULK, BACKCROSS AND SSD METHODS]

Backcross: Crossing of F_1 with either of its parents. When F_1 is crossed with homozygous recessive parent, it is called test cross.

Bulk Breeding: A selection procedure which is used in segregating population of self-pollinated species in which material is grown in bulk plot from F_2 to F_5 with or without selection, next generation is grown from bulk seed and individual plant selection is practiced in F_6 or later generations.

Donor Parent: The parent which donates desirable genes; also called non-recurrent parent, because it is used once in crossing programme.

Hybridization: Crossing between genetically dissimilar plants. It may involve two genotypes of the same species (inter-varietal hybridization) or two species of the same genus (interspecific hybridization) or two genera of the same family intergeneric hybridization).

Isogenic Lines: Genotypes having single locus difference only.

Mass Pedigree Method: A modified form of pedigree method in which segregating material is handled by bulk (mass) method when conditions are unfavourable for selection and by pedigree method when conditions are favourable for selection.

Multi-lines: The deliberate seed mixtures of isogenic lines, closely related lines or unrelated lines. A variety developed for commercial cultivation from any of these mixtures is known as multiline variety and such procedure is called multiline breeding.

Pedigree Breeding: A selection procedure which is used in segregating population of self-pollinated species and keeps proper record of plants and progeny selected in each generation.

Pedigree: Record of the ancestry of an individual selected plant for its various generations.

Recipient Parent: The parent which receives a desirable character, also called recurrent parent, because it is repeatedly used in backcrossing.

Single Seed Descent: A breeding procedure which is used with segregating population of self-pollinated species in which plants are advanced by selecting single seed per plant from F_2 generation onwards.

CHAPTER 8: BREEDING CROSS-POLLINATED CROPS [MASS SELECTION AND PROGENY SELECTION]

Line Breeding: A system of breeding in which a number of genotypes with superior performance for several characters are composed to form a variety.

Mass Selection: A method of crop improvement in which individual desirable plants are selected on the basis of phenotype from a mixed population, their seeds are bulked and used to grow the next generation **(positive mass selection).** Sometimes, only undesirable off type plants are removed from the field and rest are allowed to grow further **(negative mass selection).**

Progeny Selection: A selection procedure in which superior plants are selected from a heterogeneous population on the basis of the performance of their progeny.

CHAPTER 9: BREEDING VEGETATIVELY PROPAGATED CROPS [CLONAL SELECTION]

Clonal Selection: A procedure of selecting superior clones from the mixed population of asexually propagated crops such as sugarcane, potato, *etc.*

Clone: Progeny of a single plant obtained by asexual reproduction.

CHAPTER 10: TRANSGENIC BREEDING

Biotechnology: Application of various biological organisms or processes for mass production of useful substances or products for industry, medicine and agriculture.

Gene Revolution: Quantum jump in the productivity of various field crops through the use of transgenic cultivars.

Protoplasts: Naked cells or cells without cell wall.

Transgenes: Foreign genes or modified genes of the same species which are used for development of transgenic individuals.

Transgenic Breeding: Genetic improvement of crop plants, domestic animals and useful micro-organisms, through application of biotechnology, in relation to their economic use for mankind.

Transgenic: Individuals containing altered or transgenes from an unrelated organism. It involves taking genes from one species and inserting them into another species to get that trait expressed in the offspring.

CHAPTER 11: SMART BREEDING [MARKER ASSISTED SELECTION]

Biochemical Marker: Markers that are related to variation in proteins and amino acid banding pattern.

Cytological Marker: Markers that are related to variation in chromosome number, shape and size.

DNA Marker: A gene or other fragment of DNA whose location in the genome is known. A genetic marker is a known DNA sequence.

Epistasis: The interaction between genes. Epistasis takes place when the action of one gene is modified by one or several other genes.

Genotyping: The process of determining the genotype of an individual with a biological assay

Locus (plural loci): A fixed position on a chromosome, such as the position of a gene or a biomarker (genetic marker).

Marker assisted selection: Selection for specific alleles (which affect a trait of interest) using genetic markers.

Marker: Any genetic element (locus, allele, DNA sequence or chromosome feature) which can be readily detected by phenotype, cytological or molecular techniques, and used to follow a chromosome or chromosomal segment during genetic analysis.

Molecular Breeding: Improvement of crop plants for various economic characters through indirect selection for linked molecular markers.

Morphological Marker: In plant breeding, markers that are related to variation in shape, size, colour and surface of various plant parts.

Phenotype: Any observable characteristic of an organism, such as its morphology, development, biochemical or physiological properties, or behavior.

Pleiotropy: Influence of a single gene on multiple phenotypic traits.

Polymorphism: Presence of multiple alleles of a gene within a population, usually expressing different phenotypes

CHAPTER 12 AND 13: VARIETAL AND HYBRID SEED PRODUCTION

Breeder Seed: The progeny of nucleus seed or breeder seed produced under the strict supervision of original or sponsoring plant breeder.

Certified Seed: The progeny of either foundation or registered or certified seed. The certification is done by the State Seed Certification Agency.

Foundation Seed: The progeny of breeder seed produced by National Seeds Corporation.

Genetic Purity: The absence of seeds of other variety of the same crop as well as of other crops.

Germination: Emergence of normal seedlings from the seeds under ideal conditions of light, temperature, moisture, oxygen and nutrients.

Inert matter: Non- living materials such as sand, pebbles, soil particles, straw, *etc.*

Isolation: Separation of the field of a variety from that of another variety of the same crop to prescribed standard distance to avoid contamination.

Nucleus Seed: The initial seed of an improved variety, limited in quantity and produced by originating plant breeder.

Physical Purity: Freedom of seed from inert matter and defective seeds.

Registered Seed: The progeny of either foundation or registered seed. In India, this category is generally omitted and certified seed is produced directly from foundation seed.

Roguing: The process of removal of off type (phenotypically different) plants from the field of an improved variety to avoid contamination.

Seed Certification: A legal system which ensures production of high quality seed in terms of genetic purity and germination.

Seed testing: The process of evaluation of seeds in terms of purity and germination.

Seed: In broad sense, any plant part which is used for commercial multiplication of a crop. In strict sense, the product of fertilized ovule that consists of embryo, seed coat and cotyledon(s).

Varietal Deterioration: Permanent reduction either in the genetic or agronomic value of a released variety.

Variety: A genotype released for commercial cultivation either by State Variety Release Committee or Central Variety Release Committee.

CHAPTER 14-18: CROP BREEDING

Bagasse: In sugarcane, the fibrous residue which is left over crushing of cane.

Cereal Crops: The starchy grain of various grasses such as wheat, oats, or corn, which are used as food refer to cereal crops.

Cotton: Those species of the genus *Gossypium* which produce sinnable seed coat fibres.

Crop Plants: Domesticated and commercially cultivated plants or agricultural produce, such as grain, vegetables, or fruit, considered as a group are called crop plants: Wheat is a common crop.

Fibre Crops: Field crops which are grown for their fibers, which are traditionally used to make paper, cloth, or rope are called fibre crops. Such crops include cotton, Jute, sunnhamp, Kenef, *etc.*

Field Crop: A crop (other than fruits or vegetables) that is grown for agricultural purposes is called field crop. Cotton, pulses, sugarcane wheat, mustard, maize are examples of field crops"

Food Crops: Crops that can be used for food for humans. Examples: wheat, corn, fruits, vegetables are known as food crops.

Ginning: In cotton, the process of separating lint from the seed.

Husk: In maize, leafy structures covering the ear.

Monopodia: In cotton, vegetative branches.

Nobilization: In sugarcane, the process of crossing with *Saccharum spontaneum* and back crossing with noble cane *i.e. S. officinarum.*

Peg: In peanut, a long stalk bearing pod at the apex; also known as gynopore or carpopodium.

Pulse Crops: Leguminous plants producing edible seeds, *e.g.* chickpeas, lentils, beans, black gram, green gram, pigeon pea *etc.* are known as pulse crops.

Pulses: Those species of the family Leguminosae whose seeds are used for human consumption.

Shell: In groundnut, the pericarp of the pod.

Silk: In maize, long stigmas which emerge from upper end of cob.

Square: In cotton, unopened floral bud.

Sympodia: In cotton, fruiting branches.

Tassel: In maize, the male inflorescence.

Test Weight: Weight of 100 seeds in gram.

Trash: The dried sugarcane leaves. In cotton, dried leaf bits or any other foreign matter in the lint.

Tuber Crops: A swollen, fleshy, usually underground stem of a plant, such as the potato, bearing buds from which new plant shoots arise is known as tuber crop.

CHAPTER 19: IDEOTYPE BREEDING

Biological yield: Total dry matter production per plant.

Harvest Index: The ratio of economic yield to the biological yield or the ratio of economic produce to the total biomass.

Ideotype: A plant model which is expected to yield greater quantity of grains, fibre, oil or other useful product when developed as a cultivar.

Ideotype breeding: A method of crop improvement which is used to enhance genetic yield potential through genetic manipulation of individual plant character.

CHAPTER 20: BREEDING FOR CLIMATE CHANGE

Breeding for climate change: Plant breeding strategies used develop crop varieties for change in climate of a region.

Climate change: Any long-term significant change in the "average weather" of a region or the earth as a whole.

Weather: The day to day atmospheric conditions such as temperature, precipitation and wind patterns.

References

1. **ALLARD, R. W.** 1960. Principles of Plant Breeding. John Wiley and Sons Inc. New York.
2. **ALLARD, R. W.** 1999. Principles of Plant Breeding 2nd edition. John Wiley and Sons Inc. New York.
3. **BOROJEVIC, S.** 1990 Principles and Methods of Plant Breeding. Elsevier, Amsterdam.
4. **BRIGGS, F. N. AND KNOWLES, P. F.** 1967. Introduction to Plant Breeding. Reinhold, New York.
5. **CHOPRA, V. L. (ED.)** 2000. Plant Breeding: Theory and Practice 2nd edition, Oxford and IBH Publishing Company, Pvt. Ltd. New Delhi.
6. **FEHR, W. R.** 1987. Principles of Cultivar Development Volume 1. Theory and Techniques. Macmillan Publishing Company, New York.
7. **FREY, K. J.** (ed.) 1966. Plant Breeding. Iowa State University, Press, Ames, Iowa, USA.
8. **FREY, K. J.** (ed.) 1981. Plant Breeding II. Iowa State University, Press, Ames, Iowa, USA.
9. **HARTEN, A. V.** 1998. Mutation Breeding: Theory and Practical Applications. Cambridge Uni. UK.
10. **KUCKKUCK, H., KOBADE, G. AND WENGEL G.** 1993. Fundamentals of Plant Breeding. Narosa Publishing House, New Delhi.
11. **MAYO, O. 1984**. The Theory of Plant Breeding. Clarendon, Oxford.
12. **POEHLMAN, J. M.** 1987. Breeding Field Crops 3rd edition. AVI Publishing Co. Inc. West Port, Connecticut, USA.
13. **POEHLMAN, J. M. AND BORTHAKUR, D. N.** 1969. Breeding Asian Field Crops. Oxford and IBH Publishing Company, New Delhi.

14. **PHUNDAN SINGH** 2018. Essentials of Plant Breeding 7th edition, Kalyani Publishers, New Delhi.

15. **PHUNDAN SINGH** 2020. Principles of Plant Breeding, Kalyani Publishers, New Delhi.

16. **PHUNDAN SINGH** 2020. Genetic Principles and Plant Breeding, Brillion Publishing, New Delhi.

17. **PHUNDAN SINGH** 2020. Genetics and Plant Breeding: Comparative and Objective Analysis, Brillion Publishing, New Delhi.

18. **PHUNDAN SINGH** 2018. Plant Breeding: Molecular and New Approaches 4th edition, Kalyani Publishers, New Delhi.

19. **PHUNDAN SINGH** 2017. Fundamentals of Plant Breeding, Kalyani Publishers, New Delhi.

20. **PHUNDAN SINGH** 2018. Molecular Plant Breeding, Kalyani Publishers, New Delhi.

21. **PHUNDAN SINGH AND PRATIBHA BISEN, 2020**. Commercial Plant Breeding, Daya Publishing House, New Delhi.

22. **PHUNDAN SINGH, MONIKA SINGH AND MRIDULA BILLORE**, 2021. Commercial Plant Breeding: Objective, Daya Publishing House, New Delhi.

23. **PHUNDAN SINGH AND PRATIBHA BISEN, 2021**. Commercial Plant Breeding: At A Glance, Daya Publishing House, New Delhi.

24. **PHUNDAN SINGH, MRIDULLA BILLORE AND D. K. MISHTA, 2019**. Introduction to Maintenance Plant Breeding, Daya Publishing House, New Delhi.

25. **PHUNDAN SINGH AND RESHU TIWARI, 2021**. Introduction to Maintenance Plant Breeding: At A Glance, Daya Publishing House, NewDelhi.

26. **PHUNDAN SINGH AND D. K. MISHRA, 2021**. Introduction to Maintenance Plant Breeding: Objective, Daya Publishing House, New Delhi.

27. **PHUNDAN SINGH AND DHARMENDRA SINGH, 2021**. Introduction to Resistance Plant Breeding, Daya Publishing House, New Delhi.

28. **PHUNDAN SINGH AND DHARMENDRA SINGH, 2021**. Introduction to Resistance Plant Breeding: Objective, Daya Publishing House, New Delhi.

29. **PHUNDAN SINGH AND DHARMENDRA SINGH, 2021**. Introduction to Resistance Plant Breeding: At A Glance, Daya Publishing House, New Delhi.

30. **PHUNDAN SINGH** 2018. Concepts in Plant Breeding, Brillion Publishing House, New Delhi.

31. **PHUNDAN SINGH** 2019. Principles of crop Improvement, Daya Publishing House, New Delhi.

32. **PHUNDAN SINGH** 2019. Elements of Genetics 5th edition, Kalyani Publishers, New Delhi.

33. **PHUNDAN SINGH** 2016. Genetics 3rd edition, Kalyani Publishers, New Delhi.
34. **PHUNDAN SINGH** 2012. Cotton Breeding 3rd edition, Kalyani Publishers, New Delhi.
35. **PHUNDAN SINGH AND SANJEEV SINGH** 2010. Breeding Hybrid Cotton 3rd edn. Kalyani Publishers, New Delhi.
36. **SHARMA, J. R.** 1984. Principles and Practice of Plant Breeding. Tata McGraw-Hill Publishing Company Ltd. New Delhi.
37. **SIMMONDS, N. W.** 1979. Principles of Crop Improvement. Longman, London.
38. **SNEEP, J.AND HENDRIKSON, A. J. T.** (eds.) 1979. Plant Breeding Perspectives. Pudoc, Wageningen, Netherlands.
39. **VOSE, P. B. AND BLIXT, S. G.** (eds.) 1984. Crop Breeding A Contemporary Basis. Pergamon Press, London.
40. **WILLIAMS, W.** 1964. Genetic principles and Plant Breeding, Oxford, Blackwell.

Books by Dr. Phundan Singh

A. PLANT BREEDING

1. Essentials of Plant Breeding
2. Fundamentals of Plant Breeding
3. Plant Breeding: Molecular and New Approaches
4. Molecular Plant Breeding
5. Commercial Plant Breeding
6. Commercial Plant Breeding: At A Glance
7. Commercial Plant Breeding: Objective
8. Objective Science of Plant Breeding
9. Genetic Principles and Plant Breeding
10. Genetics and Plant Breeding: Comparative Analysis
11. Principles of Plant Breeding[For PG Students]
12. Plant Breeding: At a Glance
13. Plant Breeding [For Under Graduate Students]
14. Principles of Plant Breeding [For UG Students]
15. Practical in Crop Breeding
16. Practical and Numerical in Plant Breeding
17. Numerical Problems in Plant Breeding and Genetics
18. Glossary of Plant Breeding and Genetics
19. Breeding Crop Plants for Stress Resistance

20. Introduction to Resistance Plant Breeding
21. Resistance Plant Breeding: Objective
22. Resistance Plant Breeding: At A Glance
23. Plant Breeding: Related Legislations
24. Concepts in Plant Breeding
25. Principles of Seed Technology
26. Seed Technology: At a Glance
27. Objective Seed Technology

B. COTTON BREEDING AND GENETICS

28. Cotton Breeding
29. Heterosis Breeding in Cotton
30. Breeding Hybrid Cotton
31. Breeding Transgenic Bt. Cotton
32. Basics of Bollgard Cotton
33. Basics of Bollgard Cotton: At A Glance
34. Basics of Bollgard Cotton: Objective
35. Cotton Improvement in India
36. Sankar Kapas [In Hindi]
37. Glimpses of Cotton Breeding
38. Cotton Genetics
39. Cotton

C. GENETICS

40. Elements of Genetics
41. Genetics
42. Principles of Genetics [For UG Students]
43. Fundamentals of Genetics
44. Plant Genetics
45. Objective Genetics
46. Genetics: At a Glance
47. Objective Genetics and Plant Breeding
48. Objective Genetics and Plant Breeding [Hindi Edition]
49. Glossary cum Dictionary of Genetics and Plant Breeding
50. Molecular Genetics [Objective]
51. Molecular genetics [At a Glance]

52. Molecular Genetics [Subjective]
53. Genetics and Man
54. Genetics Today

D. QUANTITATIVE GENETICS

55. Biometrical Techniques in Plant Breeding
56. Application of Biometrical Techniques in Plant Breeding [Hindi Edition]
57. Objective Quantitative Genetics
58. Quantitative Genetics: At a Glance
59. Quantitative Genetics
60. Introduction to Agricultural Statistics
61. Agricultural Statistics: Objective
62. Agricultural Statistics: At A Glance

E. PLANT BIOTECHNOLOGY

63. Introduction to Biotechnology
64. Plant Biotechnology
65. Principles of Plant Biotechnology
66. Plant Biotechnology: At a Glance
67. Objective Plant Biotechnology
68. Fundamentals of Molecular Biology and Plant Biotechnology
69. Objective Molecular Biology and Plant Biotechnology
70. Molecular Biology and Plant Biotechnology: At A Glance

F. INTELLECTUAL PROPERTY RIGHTS

71. IPR and Plant Breeders' Rights [Subjective]
72. IPR and Plant Breeders' Rights [Objective]
73. IPR and Plant Breeders' Rights [At a Glance]
74. Introduction to Intellectual Property Rights
75. Intellectual Property Rights: At A Glance
76. Intellectual Property Rights: Objective

G. MISCELLANEOUS BOOKS

77. Elements of Baby Corn
78. Baby Corn Farming for Higher Income
79. A Guide for Competitive Examinations of Agriculture
80. Objective General Agriculture for Competitive Examinations

81. Autobiography of Phundan Singh
82. Concepts in Plant Breeding [Objective Analysis]
83. Concepts in Plant Breeding [Question and Answer]
84. Principles of Plant Breeding [Objective Analysis]
85. Principles of Plant Breeding [Question and Answer]
86. Principles of Crop Improvement [Objective Analysis]
87. Principles of Crop Improvement [Question and Answer]

Printed at Repro India Ltd.